T0181790

Lecture Notes in Networks and Systems   999

Series Editor

Janusz Kacprzyk ⓘ, *Systems Research Institute, Polish Academy of Sciences, Warsaw, Poland*

Advisory Editors

Fernando Gomide, *Department of Computer Engineering and Automation—DCA, School of Electrical and Computer Engineering—FEEC, University of Campinas—UNICAMP, São Paulo, Brazil*

Okyay Kaynak, *Department of Electrical and Electronic Engineering, Bogazici University, Istanbul, Türkiye*

Derong Liu, *Department of Electrical and Computer Engineering, University of Illinois at Chicago, Chicago, USA*

    Institute of Automation, Chinese Academy of Sciences, Beijing, USA

Witold Pedrycz, *Department of Electrical and Computer Engineering, University of Alberta, Alberta, Canada*

    Systems Research Institute, Polish Academy of Sciences, Warsaw, Canada

Marios M. Polycarpou, *Department of Electrical and Computer Engineering, KIOS Research Center for Intelligent Systems and Networks, University of Cyprus, Nicosia, Cyprus*

Imre J. Rudas, *Óbuda University, Budapest, Hungary*

Jun Wang, *Department of Computer Science, City University of Hong Kong, Kowloon, Hong Kong*

The series "Lecture Notes in Networks and Systems" publishes the latest developments in Networks and Systems—quickly, informally and with high quality. Original research reported in proceedings and post-proceedings represents the core of LNNS.

Volumes published in LNNS embrace all aspects and subfields of, as well as new challenges in, Networks and Systems.

The series contains proceedings and edited volumes in systems and networks, spanning the areas of Cyber-Physical Systems, Autonomous Systems, Sensor Networks, Control Systems, Energy Systems, Automotive Systems, Biological Systems, Vehicular Networking and Connected Vehicles, Aerospace Systems, Automation, Manufacturing, Smart Grids, Nonlinear Systems, Power Systems, Robotics, Social Systems, Economic Systems and other. Of particular value to both the contributors and the readership are the short publication timeframe and the worldwide distribution and exposure which enable both a wide and rapid dissemination of research output.

The series covers the theory, applications, and perspectives on the state of the art and future developments relevant to systems and networks, decision making, control, complex processes and related areas, as embedded in the fields of interdisciplinary and applied sciences, engineering, computer science, physics, economics, social, and life sciences, as well as the paradigms and methodologies behind them.

Indexed by SCOPUS, INSPEC, WTI Frankfurt eG, zbMATH, SCImago.

All books published in the series are submitted for consideration in Web of Science.

For proposals from Asia please contact Aninda Bose (aninda.bose@springer.com).

Nenad Filipović

Editor

Applied Artificial Intelligence 2: Medicine, Biology, Chemistry, Financial, Games, Engineering

The Second Serbian International Conference on Applied Artificial Intelligence (SICAAI)

 Springer

Editor
Nenad Filipović
Faculty of Engineering
University of Kragujevac
Kragujevac, Serbia

ISSN 2367-3370   ISSN 2367-3389 (electronic)
Lecture Notes in Networks and Systems
ISBN 978-3-031-60839-1   ISBN 978-3-031-60840-7 (eBook)
https://doi.org/10.1007/978-3-031-60840-7

This Springer imprint is published by the registered company Springer Nature Switzerland AG
The registered company address is: Gewerbestrasse 11, 6330 Cham, Switzerland

If disposing of this product, please recycle the paper.

Contents

Implementation of Hybrid ANN-GWO Algorithm for Estimation of the Fundamental Period of RC-Frame Structures

Filip Đorđević[✉] and Marko Marinković

Faculty of Civil Engineering, The University of Belgrade, Bulevar Kralja Aleksandra 73, 11000 Belgrade, Serbia
{fdjordjevic,mmarinkovic}@grf.bg.ac.rs

Abstract. The fundamental period (T_{FP}) of vibration is one of the most important parameters in structural design since it is used to assess the dynamic response of the structures. It is the time taken by a structure or system to vibrate back and forth in its most natural way, without any external forces applied. Simultaneously, T_{FP} depends on the mass distribution and stiffness of the structure, which is largely influenced by infill walls in RC frame structures, and which is why their careful design is necessary. This study aims to develop a fast, accurate, and efficient machine learning (ML) method for the prediction of the fundamental period of masonry-infilled reinforced concrete (RC) frame structures. Hybridization of the stochastic gradient descent (SGD) based artificial neural network (ANN), and meta-heuristic grey wolf optimization (GWO) algorithm is proposed as an effortless computational method. This approach provided even more reliable solutions than the robust second-order procedure based on single ML models. A total of 2178 samples of infilled RC frames were collected from available literature, where the number of storeys (*NoSt*), number of spans (*NoSp*), length of spans (*LoSp*), opening percentage (*OP*), and masonry wall stiffness (*MWS*) were considered as input parameters for predicting the output T_{FP} results. The accuracy and exploration efficiency of the proposed ANN-GWO paradigm have demonstrated superiority over existing seismic design codes and other conventional ML methods.

Keywords: Earthquake engineering · Machine learning · Artificial neural network · Grey wolf optimization · Infill frames

1 Introduction

Fundamental period of a structure is a starting point of every building design process in civil engineering. Therefore, its determination is of utmost importance. Its value determines the seismic load that will be used in the design of a building, thus it is essential to make a good estimation of its value. In the case of RC frame structures with masonry infill walls, this is a complex task. Infill walls are quite stiff and it influences the fundamental period significantly. In recent years, Artificial Intelligence (AI) techniques have been used increasingly in various engineering fields, including structural engineering [1]

N. Filipović (Ed.): AAI 2023, LNNS 999, pp. 1–6, 2024.
https://doi.org/10.1007/978-3-031-60840-7_1

and more specifically earthquake engineering [2, 3]. Existing studies are mostly based on shallow machine learning or Deep-Learning (DL) algorithms for solving regression, classification or optimization problems. This study aims to develop a fast, accurate, and efficient ML method for the prediction of the fundamental period of masonry-infilled (RC) frame structures.

2   Material and Methods

2.1   Database Description

A database consisting of 2178 masonry-infill RC frame structures was collected from the available literature FP4026 Research Database [4]. Table 1 contains information on the distribution of input and output variables, i.e. *NoSt, NoSp, LoSp, OP, MWS*, and T_{FP}. This is important because of the possibility to reproduce the results obtained in this study, but also to provide data normalization to the range $[-1 \div 1]$, which was done by using the MinMaxScaler function in the Python environment.

Table 1. Distribution description of input and output variables.

Variable	Mean	St. Dev.	Min	Max
NoSt [-]	11.50	6.35	1.00	22.00
NoSp [-]	5.76	0.87	2.00	6.00
LoSt [m]	4.77	1.45	3.00	7.50
OP [%]	31.76	28.99	0.00	75.00
MWS [10^5 kN/m]	11.38	7.85	2.25	25.00
T_{FP} [s]	0.83	0.59	0.04	3.01

Two metrics were used for the performance evaluation of the proposed methods including, mean squared error (*MSE*), and coefficient of determination (R^2) [5].

2.2   Artificial Neural Network (ANN)

ANNs are a class of machine learning models that are based on the organization and operation of the human brain. They have a number of key benefits, such as the capacity to learn from and adjust to new data, the capacity to process sizable amounts of data concurrently, and the capacity to model intricate, non-linear relationships between inputs and outputs. This study aims to predict the fundamental period of infill RC frames, using the output results from the optimization GWO technique as a starting point for ANN, in order to increase the security against falling into local minima. Validation of the results was performed by comparing the hybrid ANN-GWO model based on the SGD rule, with single first-order SGD model and second-order based adaptive moment estimation (ADAM) model developed from scratch [5]. The stability of the single model was tested using the 10-cross validation (CV) technique. The activation functions adopted for hidden and output layers are hyperbolic tangent and pure linear, respectively. SGD rule implemented in this work can be mathematically expressed as:

$$w_{i+1} = w_i - \mu \cdot g_i + \eta \cdot \Delta w_{t-1} \tag{1}$$

where $\mu = 0.1$ [0 ÷ 1], and $\eta = 0.9$ [0 ÷ 1] are learning rate and momentum as hyperparameters with values adopted according to the recommendations derived from other studies [5], g_i is the gradient calculated after each sample i, and Δw_{t-1} is the weight increment from previous iteration. Based on previous studies of the same task [6–8] where ANNs either with one or multiple hidden layers were adopted, the authors of this work investigated several architectures with a single layer and different numbers of neurons. The adopted splitting strategy considers 80% of the samples for training and the other 20% for the test phase. The most optimal ANN-GWO architecture (5–8-1) contains eight neurons in the hidden layer.

2.3 Grey Wolf Optimization (GWO)

The social structure and foraging habits of grey wolves in the wild served as inspiration for the development of the GWO meta-heuristic optimization algorithm. GWO was first proposed by [9] in 2014. As gray wolves usually live in the pack, it is important to note their social hierarchy, which consists of 4 groups: alpha (α), beta (β), delta (δ), and omega (ω). The alpha wolf as the best manager in the pack is the most dominant member, while the beta, delta, and omega wolves have a lower rank, respectively. The optimization process involves updating the positions of the wolves in the search space based on their initially random location. Grey wolves traditionally have a strict hunting procedure which can be mathematically described as follows:

$$X(t+1) = (X_1 + X_2 + X_3)/3 \tag{2}$$

$$X_1 = X_\alpha(t) - A_1 \cdot D_\alpha, X_2 = X_\beta(t) - A_2 \cdot D_\beta, X_3 = X_\delta(t) - A_3 \cdot D_\delta \tag{3}$$

$$A_i = 2 \cdot a \cdot r_1 - a, C_i = 2 \cdot r_2 \tag{4}$$

$$a = 2 \cdot (1 - iteration/maxiteration) \tag{5}$$

$$D_{\alpha,\beta,\delta} = \left| C_i \cdot X_{\alpha,\beta,\delta}(t) - X(t) \right| \tag{6}$$

where t refers to the current iteration, X is the vector of the grey wolf position, X_1, X_2, and X_3 are predicted position vectors of the alpha, beta, and delta wolves, X_α, X_β, X_δ are relative position vectors of alpha, beta, and delta wolves, A_i and C_i are coefficient vectors, r_1 and r_2 are random vectors in range [0,1], a is a component that linearly decreased from 2 to 0, and $D_{\alpha,\beta,\delta}$ are vectors that depend on the position of the prey. One of the advantages of GWO is that it has a relatively simple algorithm, without the need to calculate function derivatives, which makes it easy to implement and understand. The fundamental principle behind using GWO for training ANNs is to find the weights and biases in a more optimal manner, by reduction of the *MSE* as a fitness function between the predicted and the actual outputs. The main benefit is its superior ability to explore the high-dimensional parameter space compared to some other optimization algorithms. The authors proposed the initial random generating of 30 wolves as search agents. In the training phase, GWO iterates 10 times, followed by only 320 ANN iterations (see Table 2), which is optimal compared to at least 1000 epochs necessary for the convergence of a single conventional model from scratch. The improvement of the positions of search agents is performed by applying the procedure described in the previous section. All search agents were separately considered as vectors with 57 parameters of the neural network that were initially randomly generated in the range $[-5 \div 5]$.

3   Results and Discussion

The results presented in Fig. 1, show that in the entire domain the hybrid ANN-GWO algorithm is superior over the pure SGD and ADAM models from scratch, as well as the FEMA-450 [10] design code that proposes the following equation for calculating the fundamental period:

$$T_{FP} = 0.0466 \cdot H_n^{0.9} \tag{7}$$

where H_n is the height of the structure in meters. Table 2 summarizes the performances of the mentioned approaches, whose results are illustrated in Fig. 1. It can be concluded that the hybrid algorithm has a slightly better processing power and faster convergence (320 versus 1000 epochs) even than the second-order ADAM algorithm. A similar conclusion is reached in the case of comparison with SGD from scratch, while the robustness of the hybrid procedure is particularly pronounced in comparison to the results of the FEMA-450 design code.

Table 2. Performance indicators of the proposed hybrid and single ML models

Measure	R^2				MSE ($\cdot 10^{-3} s^2$)			
Method	ANN-GWO	SGD	ADAM	FEMA-450	ANN-GWO	SGD	ADAM	FEMA-450
Iteration	320	1000	1000	–	320	1000	1000	–
Architecture	5-8-1			–	5-8-1			–
Train	0.997	0.987	0.992	–	1.178	4.735	2.771	–
Test	0.997	0.988	0.991	–	1.189	4.894	3.194	–
All	0.997	0.987	0.992	0.562	1.179	4.767	2.856	247.690

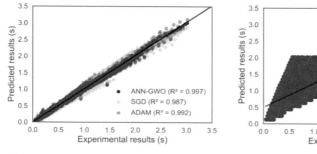
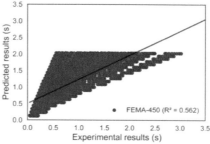

(a) ANN-GWO, SGD and ADAM models   (b) FEMA-450 design code

Fig. 1. Performance comparison of the proposed hybrid ANN-GWO model with SGD/ADAM methods, and provisions of FEMA-450 design code on all data.

4 Conclusions

This paper suggests the implementation of a meta-heuristic GWO algorithm to enhance the processing power of the ANN-SGD, used for the prediction of the fundamental period of RC frame structures. Proof that the hybrid algorithm outperforms traditional procedures is made by comparing it with the first-order ANN-SGD and the second-order ANN-ADAM algorithm made from scratch, as well as seismic design codes. ANN-GWO provided a faster convergence, and a higher coefficient of determination values, making it more accurate even than the second-order algorithm. In addition, GWO enables more successful avoidance of falling into a local minimum, which is often a problem of single ANN models. Generally, all three approaches are suitable to make predictions of the fundamental period, with some differences in speed. As part of future research, it is necessary to examine the potential application of the GWO algorithm and its hybrid variations on a wider range of problems, in order to uncover new possibilities and validate its effectiveness.

References

1. Đorđević, F.: A novel ANN technique for fast prediction of structural behavior. In: 6th International Construction Management Conference, We Build the Future (2023)
2. Rahmat, B., Afiadi, F., Joelianto, E.: Earthquake prediction system using neuro-fuzzy and extreme learning machine. In: Proceedings of the International Conference on Science and Technology (2018)
3. Celik, E., Atalay, M., Kondiloğlu, A.: The earthquake magnitude prediction used seismic time series and machine learning methods. In: IV International Energy Technologies Conference (2016)
4. Asteris, P.G.: The FP4026 research database on the fundamental period of RC infilled frame structures. Data Br., pp. 704–709 (2016). https://doi.org/10.1016/j.dib.2016.10.002
5. Đorđević, F., Marinković, M.: Advanced ANN regularization-based algorithm for prediction of the fundamental period of masonry-infilled RC frames. Adv. Comput. (2023). submitted for review

6. Asteris, P.G., et al.: Prediction of the fundamental period of infilled RC frame structures using artificial neural networks. Hindawi Publishing Corporation, Comput. Intell. Neurosci., 5104907 (2016). https://doi.org/10.1155/2016/5104907

7. Asteris, P.G., Nikoo, M.: Artificial bee colony-based neural network for the prediction of the fundamental period of infilled frame structures. Neural Comput. Appl., 4837–4847 (2019). https://doi.org/10.1007/s00521-018-03965-1

8. Tran, V.L., Kim, S.E.: Application of GMDH model for predicting the fundamental period of regular RC infilled frames. Steel Compos. Struct., 123–137 (2022). https://doi.org/10.12989/scs.2022.42.1.123

9. Mirjalili, S., Mirjalili, S.M., Lewis, A.: Grey wolf optimizer. Adv. Eng. Softw., 46–61 (2014). https://doi.org/10.1016/j.advengsoft.2013.12.007

10. FEMA-450, NEHRP recommended provisions for seismic regulations for new buildings and other structures, Part 1: Provisions, Washington (DC), Federal Emergency Management Agency (2003)

Development of a Mathematical Model for Balloon Diameter Calculation in Percutaneous Transluminal Angioplasty Using Genetic Programming

Leo Benolić[1,2(✉)]

[1] Bioengineering Research and Development Centre (BioIRC), Prvoslava Stojanovica 6, 3400 Kragujevac, Serbia
leo.benolic@gmail.com

[2] Faculty of Engineering, University of Kragujevac, Sestre Janjica 6, 34000 Kragujevac, Serbia

Abstract. This paper describes the development of a mathematical model using genetic programming to calculate the diameter of a percutaneous transluminal angioplasty (PTA) balloon dilatation catheter for a given pressure and balloon size. The dataset used for the study was provided by Boston Scientific, and the genetic programming algorithm was implemented in Python using parallel computing. The results showed high levels of accuracy, with R2 values of 0.99989 and 0.99954 for the best and parsimonious models, respectively. The developed model can be useful for *in-silico* simulations of angioplasty surgery and can contribute to improving the effectiveness of the PTA balloon dilatation catheter procedure. This study demonstrates the potential of machine learning techniques for optimizing medical device performance and design. Further work could investigate the use of other machine learning techniques and larger datasets to enhance the accuracy and generalizability of the models.

Keywords: Peripheral Artery Disease · PTA balloon · pressure · genetic programming · symbolic regression · machine learning · in-silico simulations

1 Introduction

Peripheral Artery Disease (PAD) is a pervasive vascular condition that affects millions globally. Once termed peripheral vascular disease or lower extremity arterial disease, PAD is now the universally accepted terminology. It is characterized by the narrowing or blockage of arteries, particularly in the legs, due to atherosclerotic occlusions. Systemic atherosclerosis results in these obstructions, with plaque—comprising fat, cholesterol, and calcium deposits—being the primary culprit [1–3].

Despite its prevalence, PAD often remains undetected because many patients are asymptomatic. Intermittent claudication (IC), characterized by leg pain triggered by exercise and relieved by rest, stands out as the primary symptom. Yet, not all PAD patients exhibit this symptom, which occasionally results in misdiagnoses [2, 4].

Risk factors for PAD span both lifestyle choices and genetic predispositions. These include smoking, diabetes, hypertension, and dyslipidemia. While certain risks like tobacco use and high cholesterol are modifiable, others such as age, gender, and family history are not. Given PAD's widespread impact, which varies across income levels and ethnicities, there's a pronounced need for holistic treatment solutions [2, 4].

Diagnosing PAD is streamlined with tools like the Ankle Brachial Index. Still, advanced imaging, including computed tomography and MRI, is essential for detailed interventional planning. Treatments have evolved over time, focusing on enhancing patient quality of life and minimizing vascular complications. Interventional approaches, particularly percutaneous transluminal angioplasty (PTA), have seen remarkable advancements [1, 3].

PTA is a minimally invasive technique tailored for PAD patients, notably those with intermittent claudication or critical limb ischemia. It boasts a low complication rate and a high success rate. Yet, a significant challenge remains its restenosis rate. Stenting, especially using nitinol self-expanding stents, has risen as a robust alternative, especially for challenging cases. Such stents are known for their radial strength, shape-memory attributes, and minimal foreshortening. Research, including the Dutch Iliac Stent Trial and the Zilver PTX randomized trial, highlights stenting's efficacy, sometimes surpassing PTA. Consequently, while PTA is foundational for treating PAD, modern stenting techniques have enriched the treatment landscape, ensuring improved patient outcomes [5].

Machine learning (ML), a subset of artificial intelligence (AI), has been progressively integrated into various facets of medicine, manifesting its transformative potential. For instance, the Neuro Fuzzy hybrid model (NFHM) provides a cutting-edge approach in blood pressure classification by assimilating neural networks and fuzzy logic, demonstrating an aptitude for accurate hypertension and hypotension diagnoses [6]. Concurrently, in the realm of imaging, 3D convolutional neural networks are being employed to discern types of carotid arteries, albeit with a cautionary note on the vulnerabilities inherent to deep learning models [7]. In another domain, femoral peripheral artery disease (PAD) treatment, an advanced decision-support system has been pioneered using the radial basis function neural network (RBFNN). This system not only exhibited superiority over conventional models but also elucidated the impressive strides AI is making in augmenting clinical decision-making [8]. Complementing these advancements, Mistelbauer et al. unveiled a semi-automatic technique adept at identifying vessels in challenging PAD scenarios, reinforcing the versatility and efficacy of ML-driven methods across diverse medical applications [9]. Collectively, these studies underscore the burgeoning role and invaluable contributions of ML algorithms in the expansive landscape of medical diagnostics and therapeutics.

Genetic programming (GP), a specialized branch of artificial intelligence, has seen remarkable applications across the vast medical landscape. A study by [10] illustrated GP's capability in epidemiology, where it was employed to derive symbolic expressions that accurately estimated the epidemiology curve for the COVID-19 pandemic in the U.S. Similarly, GP's potential in radiology was demonstrated in study [11], where it aided in interpreting cervical spine MRI images, reaching a remarkable prediction accuracy of 90%. Beyond radiology, the method has also found utility in neurology, with research [12] utilizing Cartesian Genetic Programming to detect subtle symptoms of Parkinson's disease by analyzing patients' pen movements during a figure copying test. In the realm of bioinformatics, Hu [13] employed Linear Genetic Programming to evolve symbolic models predicting disease risks based on metabolite concentrations in blood, showcasing GP's intrinsic feature selection capability. Lastly, Bartsch [14] harnessed GP in oncology, developing classifiers to predict recurrence in Nonmuscle-Invasive Urothelial Carcinoma based on gene expression profiling. These diverse studies exemplify GP's profound versatility and potential in reshaping diagnostic, predictive, and epidemiological paradigms in medicine.

In the realm of angioplasty, the precise estimation of a balloon's current diameter based on its nominal diameter and pressure is paramount. The overarching aim of this study is to develop a mathematical model that can accurately predict the current diameter of an angioplasty balloon given its nominal diameter and the pressure applied.

The rationale behind striving for a mathematical model as opposed to a software-centric solution is multifold. Firstly, mathematical models offer universal applicability, uninfluenced by the constraints of programming languages. Secondly, they are immune to the limitations imposed by specific operating systems or hardware configurations. This transcends the barriers of technological specificity, allowing for a wider spectrum of usability.

The choice of Genetic Programming (GP) as the methodological backbone for this endeavor is grounded in its unique capabilities. Genetic Programming, rooted in the principles of evolutionary processes, possesses the ability to craft mathematical formulas using predefined structures. Instead of manually searching or tweaking parameters, GP autonomously evolves the optimal formula that best represents the underlying relationship between the variables. This evolution-inspired approach aids in circumventing the challenges associated with traditional mathematical modeling, delivering a solution that is both accurate and robust.

In light of these considerations, this study hypothesizes that through the application of Genetic Programming, we can derive a universal formula that, when fed with the nominal diameter and current pressure of a balloon, can predict its current diameter with high precision. This model can be useful for *in-silico* simulations of angioplasty surgery.

2 Materials and Methods

In this study, we utilized a dataset provided by the manufacturer Boston Scientific [15], which contains the diameter to pressure ratio for various sizes of the PTA balloon dilatation catheter. The dataset comprises balloon diameters ranging from 3 mm to 14 mm and various lengths, spanning from 40 mm to over 130 mm. During preprocessing, it was observed that for some balloons, the actual diameter at nominal pressure might slightly deviate from the designated diameter. For instance, a balloon labeled as 12 mm might not exactly measure this value at its nominal pressure. For the purpose of model approximation, our dataset was divided into two segments: 70% for training and 30% for validation/testing.

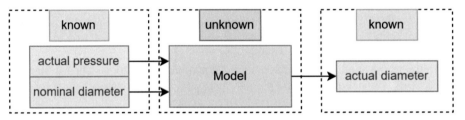

Fig. 1. Black-box model

In the evolving domain of machine learning, algorithms operate on the principle of learning patterns and relationships from data without explicit programming. Such algorithms can often be visualized as a 'black box' where inputs are transformed into outputs based on learned patterns, but the exact inner workings and relationships might not be immediately clear. Figure 1 showcases a typical black box model with two inputs, 'actual pressure' and 'nominal diameter', producing an output, 'actual diameter'.

While this black box representation simplifies the understanding of machine learning models, it is crucial to have methods that can deduce the nature of relationships inside this box. This is where Genetic Programming (GP) comes into play. GP, as an algorithmic approach, evolves and finds the best-fitted formula or relationship to map given inputs to outputs. By using GP, we can potentially unravel the 'equations' or 'formulas' governing the black box, offering us a more transparent view of how inputs are being transformed. This interpretability is especially crucial in fields like medicine, where understanding the rationale behind predictions is as important as the accuracy of the prediction itself.

Genetic Programming (GP) is a machine learning technique that draws inspiration from Charles Darwin's principles of natural evolution. It is a part of the broader family of genetic algorithms, which are metaheuristic methods that mimic the processes of evolution, including mutation, crossover, and reproduction.

In this study, the GP technique was applied using symbolic regression to derive mathematical models from the dataset. GP essentially creates a symbolic mathematical model visualized as a tree. Within this tree, the nodes represent various mathematical functions, while the leaves are either specific variables, such as X0, X1, X2, or constants. To provide a clearer picture, an equation like y = X0 × 3.4 + sin(X1) can be depicted as a tree, as demonstrated in Fig. 2.

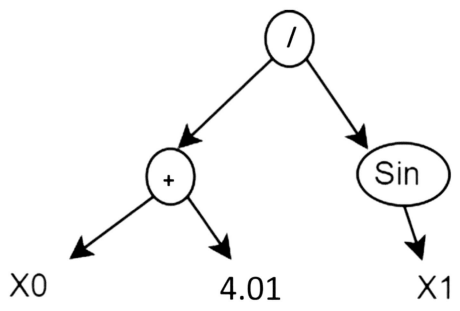

Fig. 2. Example of an equation tree

These mathematical trees are initially established through a process known as "Random Initialization", where the initial population is formed at random. Following this, parents are selected from this population based on specific fitness criteria. Once parents are determined, they undergo a series of genetic operations, with the offspring being produced through processes like mutation, crossover, and reproduction. This new generation, known as the "Offspring", is subsequently evaluated for their accuracy and relevance.

From this group, trees that demonstrate exceptional performance are chosen for the next cycle using a technique termed "Tournament Selection". These top-performing trees are the "Survivors" of this round. The algorithm evaluates whether to continue or halt based on certain "Stopping Criteria". If the criteria dictate continuation, the cycle loops back to the parent selection and proceeds as before. If the criteria suggest termination, the process ends. This cycle of evolution persists until a preset number of iterations are reached or other stipulated stopping points are met. A comprehensive depiction of the Genetic Programming process can be observed in Fig. 3.

In order to adjust the behavior of the algorithm, it is necessary to define the hyper-parameter rules or at least their range (Table 1).

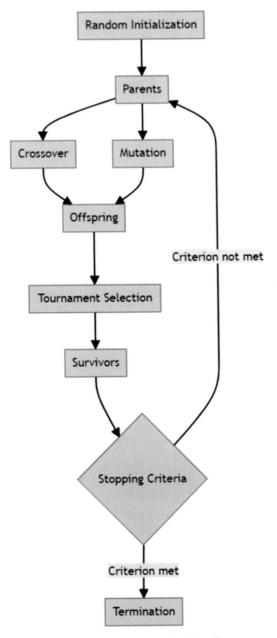

Fig. 3. Example of GP algorithm flow

Model quality estimation:
To gauge the effectiveness of the resulting model, we employ the coefficient of determination, R2. This metric serves as an indicator of how accurately the statistical model aligns with the observed data. Essentially, R2 represents the fraction of variance in the

Table 1. Hyper parameters

	Parameter range	
	min	max
Population	100	550
Generation	150	650
Tournament size	5	20
Crossover probability	0.9	0.97
Subtree mutation probability	0	0.09
Hoist mutation probability	0.01	0.07
Point mutation probability	0	0.09
Sum of probability	0.9	1
Constant range	−5000	5000
The minimal initial tree depth	3	8
The maximal initial tree depth	11	11
Stopping criteria coefficient	0,00001	
Parsimony coefficient	0.001	0.1
Metrics	RMSE	
Function set	add, sub, mul, div, sqrt, log, abs, neg, inv, max, min, sin, cos, tan	

dependent variable captured by the model. The Equation:

$$R^2 = 1 - \frac{SSR}{SST} \tag{1}$$

where: SSR is a Sum of Squared Regression (variation explained by model), and SST is Sum of Squared Total (total variation in data) [17].

3   Results and Discussion

In our exploration of the genetic programming algorithm's application, it became evident that computational efficiency is paramount, especially when dealing with complex datasets and the necessity for multiple algorithm runs. Consequently, the approach we took, as visualized in our diagram (refer to Fig. 4), centered around parallelizing our computations.

The decision to implement our algorithm in Python was bolstered by the availability and robustness of the gplearn 0.4.2 library. It provided us with a solid foundation to build our genetic programming routines. However, as with any computationally intensive task, optimization of runtime was a key challenge.

To address the efficiency concern, we integrated the multiprocessing library, a staple in Python's parallel computing arsenal. By deploying our GP algorithm concurrently on four separate processor cores, we were not only able to maximize our computer's resource utilization but also significantly cut down on training durations. This parallel execution method was visually depicted in a comprehensive diagram, which can be found in the Fig. 4 of this paper.

The system's unique design allowed each core to execute the GP algorithm independently, each initializing with distinct random states and parameters. This strategy proved invaluable, especially when it came to hyperparameter tuning. By repeatedly running the GP algorithm in parallel, we were granted multiple opportunities to refine our hyperparameters in a fraction of the time it would have taken on a single-core execution. This parallel computing approach, as facilitated by Python's multiprocessing library, has proven instrumental in enhancing the efficiency of our genetic programming endeavors. It underscores the importance of leveraging modern computing architectures and libraries to overcome challenges inherent in machine learning and data processing tasks.

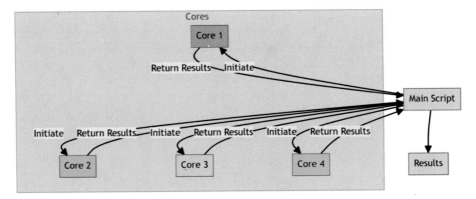

Fig. 4. Example of Multiprocessing Utilization

The results of the study show the curves of the obtained models for 5 mm and 3 mm balloons, as presented in Figs. 1 and 2, respectively. The graphs indicate the actual data points provided by the manufacturer. The deviations observed in the figures are less than 40 microns, which is less than 2% deviation. The accuracy of the models is high with R2 $= 0.99989$. These results demonstrate the effectiveness of using the genetic programming approach for developing the mathematical model for calculating the balloon diameter for a given pressure and balloon size. The developed model can be used for *in-silico* simulations of angioplasty surgery and can contribute to improving the effectiveness of the PTA balloon dilatation catheter procedure. However, further validation of the model using clinical data is necessary to assess its clinical applicability.

In the discussion, it can be explained that Figs. 5 and 6 show the syntax tree of the best performing model with an R2 of 0.99989 and a slightly inferior model with an R2 of 0.99954, respectively. The difference in performance was achieved by changing the parsimony coefficient, which penalizes models with more parameters during the evaluation process. This was done to avoid increasing the size of the models without any noticeable improvement in accuracy. The use of the parsimony coefficient helped to

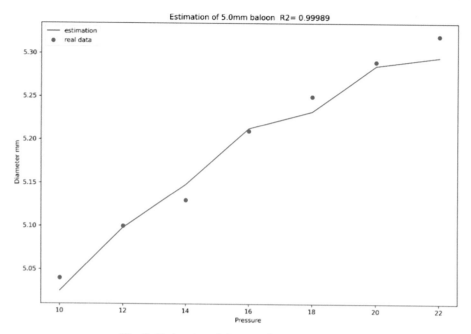

Fig. 5. Estimation of diameter for 5 mm Balloon

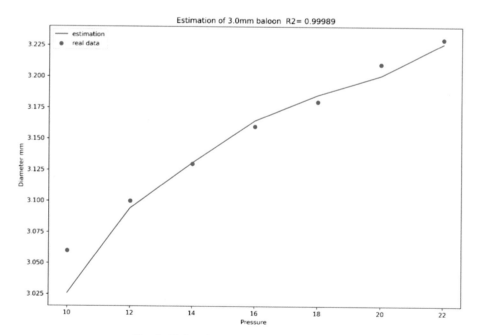

Fig. 6. Estimation diameter for 3 mm Ballon

ensure that the models remained computationally efficient while achieving high levels of accuracy (Fig. 7).

Fig. 7. Syntax tree of the best performing model with an R2 of 0.99989

An interesting feature of the model with an R^2 of 0.99954 (Fig. 8) is its reduced tree size. Although its performance is marginally lower, its simplicity is commendable. Indeed, the formula governing this model is so straightforward that it can be computed even with a basic calculator:

$$= X1 * \frac{sinsin\left(\sqrt{X0}\right) * log(X1^{-1})}{\sqrt{X0} - 48.522} = X1 * \frac{sinsin\left(\sqrt{X0}\right) * log(X1^{-1})}{\sqrt{X0} - 48.522} \qquad (2)$$

This highlights one of the prime advantages of employing genetic programming. While complex models can achieve remarkable accuracy, simpler models derived from the

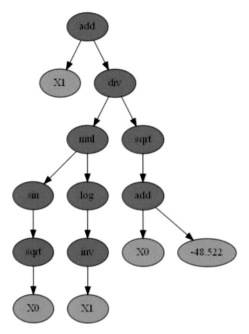

Fig. 8. Syntax tree of smaller model with an R2 of 0.99954

same methodology can be equally valuable, especially when the ease of calculation or resource constraints are considered.

Figures 9 and 10 expand on the applicability of our derived model to a broader range of balloon diameters. Figure 9 specifically showcases the estimated curves for diameters ranging from 3 mm to 8 mm, while Fig. 10 emphasizes on the larger spectrum, depicting diameters from 9 mm to 14 mm.

What is particularly noteworthy here is the inclusion of several non-standard diameters in both figures. These non-standard sizes, often overlooked in conventional representations due to their irregularity or rarity, have been effortlessly predicted using our genetic programming-derived model. The ability to forecast such unconventional diameters exemplifies the robustness and adaptability of our model, making it a comprehensive tool for diverse applications.

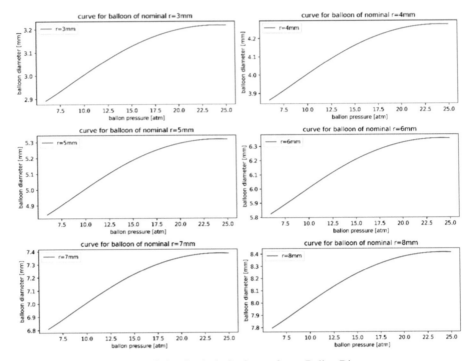

Fig. 9. Predictive Analysis for 3 mm–8 mm Ballon Diameters

These figures underscore the potential of our approach not only in accommodating standardized medical measurements but also in pioneering predictions for novel or less common balloon sizes. This adaptability can be crucial in specialized medical scenarios or research applications where non-standard sizes are required.

In essence, Figs. 9 and 10 illustrate the expansive capability of our model, emphasizing its value not just for conventional, but also for avant-garde applications in the realm of angioplasty.

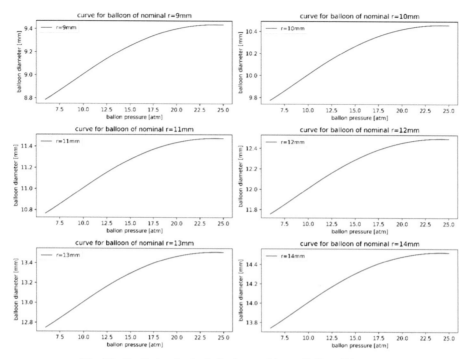

Fig. 10. Predictive Analysis for 9 mm–14 mm Balloon Diameters.

4   Conclusions

In conclusion, we have successfully applied machine learning techniques, specifically genetic programming, to model the relationship between balloon diameter and pressure in a balloon catheter system. By using a dataset provided by the manufacturer, we were able to train and evaluate the models, achieving high levels of accuracy with $R2$ values of 0.99989 and 0.99954 for the best and parsimonious models, respectively. These models were able to predict balloon pressure with an accuracy of less than 2% deviation from the actual data. The parallel computing approach, implemented using the multiprocessing library in Python, allowed us to reduce the training time and maximize the utilization of computing resources. The results demonstrate the potential of machine learning techniques for optimizing medical device performance and design. Further work could investigate the use of other machine learning techniques and larger datasets to enhance the accuracy and generalizability of the models.

Acknowledgement. This research is supported by the DECODE project that has received funding from the European Union's Horizon 2020 research and innovation programme under the Marie Skłodowska-Curie grant agreement No 956470. This article reflects only the author's view. The Commission is not responsible for any use that may be made of the information it contains.

References

1. Conte, S.M., Vale, P.R.: Peripheral arterial disease. Heart Lung Circ. **27**(4), 427–432 (2018)
2. Criqui, M.H., Aboyans, V.: Epidemiology of peripheral artery disease. Circ. Res. **116**(9), 1509–1526 (2015)
3. Remarkable care, UW Health. https://patient.uwhealth.org/healthfacts/7570. Accessed 1 Mar 2023
4. Benjamin, E.J., Muntner, P., Alonso, A., Bittencourt, M.S., Callaway, C.W., Carson, A.P.: American heart association council on epidemiology and prevention statistics committee and stroke statistics subcommittee. Heart disease and stroke statistics—2019 update: a report from the American Heart Association. Circulation **139**(10), e56-e528 (2019)
5. Schillinger, M., Minar, E.: Percutaneous treatment of peripheral artery disease: novel techniques. Circulation **126**(20), 2433–2440 (2012)
6. Melin, P., Guzman, J., Prado-Arechiga, G.: Artificial intelligence utilizing neuro-fuzzy hybrid model for the classification of blood pressure. J. Hypertens. **34**, e162–e163 (2016)
7. Le, E., Tarkin, J., Evans, N., Chowdhury, M., Rudd, J.: 875 Using stress testing to identify vulnerabilities in artificial intelligence models for the identification of culprit carotid lesions in cerebrovascular events. British J. Surg. **108**(Supplement_6), znab259–1123 (2021)
8. Yurtkuran, A., Tok, M., Emel, E.: A clinical decision support system for femoral peripheral arterial disease treatment. Comput. Math. Methods Med. 2013 (2013)
9. Mistelbauer, G., et al.: Semi-automatic vessel detection for challenging cases of peripheral arterial disease. Comput. Biol. Med. **133**, 104344 (2021)
10. Anđelić, N., et al.: Estimation of COVID-19 epidemiology curve of the United States using genetic programming algorithm. Int. J. Environ. Res. Public Health **18**(3), 959 (2021)
11. Wang, C.S., Juan, C.J., Lin, T.Y., Yeh, C.C., Chiang, S.Y.: Prediction model of cervical spine disease established by genetic programming. In: Proceedings of the 4th Multidisciplinary International Social Networks Conference, pp. 1–6 (2017)
12. Smith, S.L., Gaughan, P., Halliday, D.M., Ju, Q., Aly, N.M., Playfer, J.R.: Diagnosis of Parkinson's disease using evolutionary algorithms. Genet. Program Evolvable Mach. **8**, 433–447 (2007)
13. Hu, T.: Can genetic programming perform explainable machine learning for bioinformatics?. Genetic Programming Theory and Practice XVII, 63–77 (2020)
14. Bartsch, G., et al.: Use of artificial intelligence and machine learning algorithms with gene expression profiling to predict recurrent nonmuscle invasive urothelial carcinoma of the bladder. J. Urol. **195**(2), 493–498 (2016)
15. https://www.bostonscientific.com/en-US/products/catheters--balloon/Mustang_Balloon_Dilatation_Catheter.html. Accessed 1 Mar 2023
16. De Jong, K.: Learning with genetic algorithms: an overview. Mach. Learn. **3**(2), 121–138 (1988)
17. Turney, S.: Coefficient of Determination (R^2) | Calculation & Interpretation. Scribbr. Retrieved May 3, 2022 (2022). https://www.scribbr.com/statistics/coefficient-of-determination/

Estimation of Oscillatory Comfort During Vertical Vibrations Using an Artificial Neural Network

Igor Saveljic[1,2]([✉]), Slavica Macuzic Saveljic[3], Branko Arsic[2,4], and Nenad Filipović[3]

[1] Institute for Information Technologies, University of Kragujevac, Jovana Cvijića Bb, 34000 Kragujevac, Serbia
isaveljic@kg.ac.rs
[2] Bioengineering Research and Development Center, Prvoslava Stojanovica 6, 34000 Kragujevac, Serbia
[3] Faculty of Engineering, University of Kragujevac, Sestre Janjic 6, 34000 Kragujevac, Serbia
[4] Faculty of Science, Department of Mathematics and Informatics, University of Kragujevac, Radoja Domanovica 12, 34000 Kragujevac, Serbia

Abstract. Vertical vibrations in the vehicle can have a negative effect on the oscillatory comfort of the passengers. Vibrations can cause pain in muscles and joints, fatigue, problems with vision and coordination of movements. The ISO 2631 standard provides guidelines for assessing the level of vibration experienced by passengers in a vehicle and provides guidelines for the design and testing of vehicles to ensure the highest possible oscillatory comfort. In this paper, the influence of vertical vibrations on the human body was investigated using the ISO 2631 standard. For the purposes of the experiment, ten healthy subjects participated, five male and five female. Based on experimental results, an artificial neural network was developed to evaluate oscillatory comfort.

Keywords: vertical vibrations · oscillatory comfort · passengers · artificial neural network

1 Introduction

Whole-body vibrations represent one of the indicators of the improvement of the human body's kinematic part. Long exposure of the body to vertical vibrations can lead to harmful consequences on the musculoskeletal system [1]. When analyzing vibrations, it is important to fully describe the characteristic values of vibrations. For the human body exposed to vibrations, the following characteristic values are important: the frequency spectrum, amplitude, action factor, and length of exposure to vibrations [2].

Exposure to vibrations has a different impact on the human body, ranging from minor discomfort to decreased work efficiency and health disturbances. Today, there are guidelines in the international standard ISO 2631–1:1997 [3] for defining the tolerance of the human body exposed to whole-body vibrations. The standard ISO 2631-1:1997 is used for assessing exposure to high levels of vibrations and impacts. Vibrations that the human body absorbs can lead to muscle contractions that can cause muscle fatigue, especially at resonant frequencies. Vertical vibrations in the range of 5 Hz–10 Hz cause resonance in the thorax - abdominal system (4 Hz–8 Hz in the chest, 20 Hz–30 Hz in the head, neck, and shoulders and 60 Hz–90 Hz in the eyes) [3].

The ISO 2631–1:1997 standard provides acceptable values of human discomfort depending on daily exposure to vibrations, but does not define specific limits. In Table 1, the mentioned values of discomfort are shown.

Table 1. Passenger comfort perception according to ISO 2631 (1997)

Acceleration (m/s^2)	Comfort Level
<0.315	Not uncomfortable
0.315–0.63	A little uncomfortable
0.5–1	Fairly uncomfortable
0.8–1.6	Uncomfortable
1.25–2.5	Very uncomfortable

Also, the SRPS ISO 2631–1:2014 [4] standard specifies the total value of the average effective rms acceleration, based on which the assessment of the impact on discomfort is carried out according to the formula:

$$a_v = \sqrt{\left(k_x \cdot \ddot{x}_{rms,w}\right)^2 + \left(k_y \cdot \ddot{y}_{rms,w}\right)^2 + \left(k_z \cdot \ddot{z}_{rms,w}\right)^2} \qquad (1)$$

where: k_x, k_y, k_z – are correction factors for r.m.s. values of weighted accelerations in the direction of the x, y and z axis,– x, z and z with two poinst above are mean effective value of the weighted acceleration for the directions x, y and z.

The development of new technologies, specifically artificial intelligence, provides the opportunity for a better understanding of human body behavior exposed to vibrations [6]. The term "artificial intelligence" refers to an inanimate system that demonstrates the ability to adapt to new situations. It is based on the behavior of human beings, in order to fully replicate human behavior [5]. Machine learning is a subset of artificial intelligence that deals with building computer systems that can adapt to new situations. This field has been extremely popular in recent years, both in science and industry [9].

In this work, an ANN model was created based on the experimental results. Based on this, it is possible to perform an assessment of comfort in the vehicle.

2   Materials and Methods

The research on the impact of vibrations on humans is conducted with two aspects [10, 11]:

- health (fatigue, oscillatory discomfort, appearance of professional improvements),
- mechanical frequency response of the human body (biodynamics).

In literature, the impact of harmonic and stochastic vibrations on humans is most often considered. In these cases, the frequency range is in the range of 1 Hz–30 Hz. Recent research has shown that humans are also very sensitive to frequencies below 1 Hz [12]. For these research, or measuring the impact of vibrations on humans, vibration exciters, or shakers, are used. They are most commonly realized on the hydraulic principle, as a hydraulic platform with the possibility of exciting in two independent axes (two-axis) or in one linear axis (one-axis).

For this research, a hydro-dynamic shaker, HP-2007, was used. The HP-2007 hydro-dynamic shaker is designed to provide excitation in the frequency range of 0.1 Hz–31 Hz and amplitude of 0 mm–50 mm.

In the study, 10 subjects participated (5 males and 5 females). The average values for five healthy male subjects were: 30.8 years of age, 183.6 cm height, 90.2 cm seating height, 93.4 kg weight, and 27.7 BMI. Also in the experiment, five healthy female subjects with average age of 30.4, height of 172.6 cm, seating height 79.2, weight 68.4 kg and BMI 22.9 participated. Both male and female subjects were exposed to random vertical vibrations for one excitation value of 0.45 m/s^2 r.m.s. in the frequency range of 0.1 Hz–20 Hz. The angle of inclination of the seat was 90° (Fig. 1).

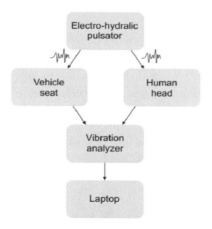

Fig. 1. A laboratory setup of the experiment

After measuring the experimental values of ponderous acceleration on the subjects, a neural network was used for model formation. Neural networks can be described as a relatively new concept used in data analysis. Their wide application can be seen in social and technical sciences as well as many other fields.

In this work, a common artificial neural network was used to determine the oscillatory comfort in a vehicle. The neural network is defined with 50 neurons in the first hidden layer and one neuron in the output layer. The data used for training the network was data obtained experimentally from the subjects. For each subject, the following parameters were given: BMI, height, weight, seat height, age, gender, years and measured r.m.s values.

Figure 2 shows the schema of the artificial neural network used in this work.

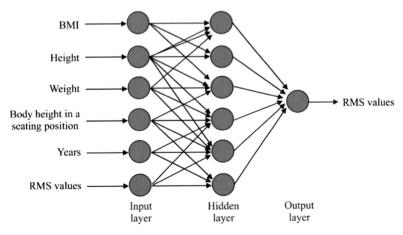

Fig. 2. A scheme of a neural network

The mean absolute error is used as the loss function in the training phase. The minimization of the loss function is done using the Adam optimization algorithm.

3   Results

3.1   Experimental Results

Based on the ISO 2631 standard, this study carried out an assessment of oscillatory comfort based on the overall effective comfort of the weighted acceleration. Table 2 shows the overall r.m.s. values of weighted acceleration and the assessment of oscillatory comfort under the influence of vertical motion for 5 male test subjects. Minor discomfort is observed for test subjects numbered 2 and 3, while no discomfort was observed in the other test subjects.

Table 2. R.m.s. weighted acceleration values and oscillatory comfort ratings under the influence of vertical excitation for five male subjects

ID	r.m.s	Comfort level
1	0.364	A little uncomfortable
2	0.352	A little uncomfortable
3	0.179	Not uncomfortable
4	0.186	Not uncomfortable
5	0.284	Not uncomfortable

In Table 3, the overall r.m.s. values of the weighted acceleration and the assessment of oscillatory comfort under the influence of vertical vibration are shown for 5 female test subjects.

Table 3. R.m.s. weighted acceleration values and oscillatory comfort ratings under the influence of vertical excitation for five female subjects

ID	r.m.s	Comfort level
1	0.396	A little uncomfortable
2	0.297	Not uncomfortable
3	0.376	A little uncomfortable
4	0.355	A little uncomfortable
5	0.288	Not uncomfortable

Based on the Table 3, it can be concluded that there is a little discomfort observed for the subjects under the serial number 1, 3 and 4, while no discomfort was observed in the other two subjects. From Table 2 and 3, differences between male and female subjects can be observed and it can be said that the testing conditions are more suitable for the male population than the female.

3.2 Artificial Neural Network Results

Using artificial neural network and ISO standard 2631, it is possible to train the neural network. The ANN model is trained with 100 training epochs. The coefficient in training, validation and testing was 92.1% for male subjects. The mean square error for the predicted RMS values of male subjects was 0.051. In the female population, the coefficient in training, validation and testing was 91.3%. The mean square error for the predicted RMS values of female subjects was 0.069. It is known that there are differences in the anatomy of both sexes, but it should be noted that the percentage of visceral fat is higher in the female population than in the male population. Special analysis were made for male and female subjects due to the different anthropometric characteristics of both sexes.

In Tables 4 and 5, the predicted RMS values of male subject under serial number 5 and female subject under serial number 5 are shown under the action of vertical vibration.

Table 4. Predicted r.m.s values for male subject number 5

ID	Original	Predicted
R.m.s	0.284	0.269
Comfort level	Not uncomfortable	Not uncomfortable

The advantage of applying neural networks can be seen based on these results. With a large amount of experimental data, the designed neural network can provide a good basis for further analysis of vehicle comfort and replace expensive and time-consuming experiments.

Table 5. Predicted r.m.s values for female subject number 5

ID	Original	Predicted
R.m.s	0.288	0.298
Comfort level	Not uncomfortable	Not uncomfortable

4 Conclusions

In this study, the assessment of the vehicle's oscillatory comfort was carried out based on the ISO 2631 standard. Ten test subjects were exposed to random single-axis vertical vibrations. Five healthy male and five healthy female test subjects participated in the experiment. It was determined that there are small differences in the comfort assessment between males and females. The excitation of 0.45 m/s^2 r.m.s. showed greater discomfort among females than among males. The reason for this may be differences in anthropometric characteristics of the test subjects.

The artificial neural network model developed in this study shows the ability for precise biomodeling. The main feature of this model was to take into account the input data of the test subject, namely, height, weight, sitting height, BMI, and years during exposure to whole body vibrations.

The continuation of this research would be in the direction of determining oscillatory comfort in relation to changes in seat inclination angle and changes in excitation values, which will be future research.

Acknowledgement. This research was supported by the Ministry of Education, Science and Technological Development of the Republic of Serbia through Grant TR35041 and 451–03-47/2023–01/200378. Also, this research is supported by the European Union's Horizon 2020 research and innovation programme under grant agreement No 952603. This article reflects only the author's view. The Commission is not responsible for any use that may be made of the information it contains.

References

1. Nawayseh, N., Griffin M.J.: Tri-axial forces at the seat and backrest during whole-body vertical vibration. J. Sound Vib., 309–326 (2004)
2. Rakheja, S., Dong, R.G., Patra, S., Boileau, P.-É., Marcotte, P., Warren, C.: Biodynamics of the human body under whole-body vibration: Synthesis of the reported dataInternational. J. Ind. Ergon. **40**, 710–732 (2010)
3. ISO 2631–1, Evaluation of Human Exposure to Whole-Body Vibration. Part 1: General Requirements, International Organization for Standardization Geneva (1997)
4. ISO 2631–1, Mehaničke vibracije i udari - Vrednovanje izlaganja ljudi vibracijama celog tela - Deo 1: Opšti zahtevi (2014)
5. Eletter, S.F., Yaseen, S.G., Elrefae, G.A.: Neuro-based artificial intelligence model for loan decisions. Am. J. Econ. Bus. Adm. **2**(1), 27–34 (2010)
6. Won, S.H., Song, L., Lee, S.Y., Park, C.H.: Identification of finite state automata with a class of recurrent neural networks Neural Net. IEEE Trans. **21**, 1408–1421 (2010)

7. May, P., Zhou, E., Lee, C.W.: Learning in fully recurrent neural networks by approaching tangent planes to constraint surfaces. Neural Net. **34**, 72–79 (2012)
8. Saveljic, I., Saveljic, S.M., Filipovic, N.: Numerical analysis of vibration effects on the lumbar spine. In: 20th International Symposium Infoteh-Jahorina (2021)
9. Saveljic, S.M., Arsic, B., Saveljic, I., Lukic, J.: In-vehicle comfort assessment during fore-and-aft random vibrations based on artificial neural networks (ANN). In: IX International Congress Motor Vehicles and Motors (MVM 2022). IOP Conference Series: Materials Science and Engineering, vol. 1271, pp. 1–7 (2022)
10. Mansfield, N.J.: Human response to vehicle vibration. In: Automotive Ergonomics: Driver-vehicle interaction, pp. 77–96. CRC Press, Florida (2013)
11. Toward, M.G.R.T., Griffin, M.J.: Apparent mass of the human body in the vertical vibration: effect of seat backrest. J. Sound Vib. **327**, 657–669 (2009)
12. Gan, Z., Hillis, A.J., Darling, J.: Development of a biodynamic model of a seated Human body exposed to low frequency whole-body Vibration. In: 11th International Conference on Vibration Problems (2013)

HR Analytics: Opportunities and Challenges for Implementation in SME

Zeljko Z. Bolbotinovic[1]([✉]), Ranka M. Popovac[2], Dragan V. Vukmirovic[3], Tijana Z. Comic[1], and Nebojsa D. Stanojevic[3]

[1] Tekijanka D.O.O, Dunavska 50, 19320 Kladovo, Serbia
`{zeljko,tijana}@tekijanka.com`
[2] Philip Morris International, Lausanne, Switzerland
`ranka.popovac@pmi.com`
[3] Faculty of Organizational Science, Jove Ilica 154, 11000 Belgrade, Serbia
`dragan.vukmirovic@fon.bg.ac.rs, ns20205003@student.fon.bg.ac.rs`

Abstract. The paper explores opportunities and challenges for implementing human resourcing analytics (HRa) in small and medium enterprises (SMEs) in the first stages of their digital transformation. In the reference literature, this topic is not sufficiently represented, regardless of the importance that SMEs have in the economic system. The solutions presented and analyzed are primarily related to the digital transformation of large companies, which have much more resources at their disposal than SMEs, so these solutions are largely inapplicable to SMEs. For this purpose, an original model for the beginning of digital transformation was defined, which envisages the implementation of HR in SMEs (HRPA - HR People Analytic model). The main components of the model are presented and directions for implementation are indicated. The methodological procedure applied in the research is based on the literature review, the analysis of secondary sources and the results of the author's previous research. Initial hypotheses were set, and recommendations were made for further development of the model.

Keywords: HR analytics · people analytics · SMEs · digital transformation · model · implementation

1 Introduction

SMEs are the backbone of many countries' economies and important contributors to job creation and global economic development. In Europe alone, before the Covid-19 pandemic existed 24 million small businesses, which employed a total of 95 million people and generated 4 trillion (10^{12}) euros per year [1].

Research gap: After the literature review, we concluded that regardless of the dominant role of SMEs across the globe, they remain dramatically underrepresented in academic research [2]. HRa shares this fate. Even considering SMEs, one of the crucial problems is that many studies "have perpetuated a large firm bias, by either uncritically deploying established research instruments, and/or by casting the small firm as

N. Filipović (Ed.): AAI 2023, LNNS 999, pp. 28–32, 2024.
https://doi.org/10.1007/978-3-031-60840-7_4

lacking or deficient if they fail to meet normative ideals" [2]. To overcome the digital divide between large companies and SMEs, we proposed a segmented approach to digital transformation [3] of SMEs that starts with one business segment (function or process). Based on the results of previous research [4], we decided to start by digitizing the human resource (HR) business function by introducing HRa.

For this paper, we defined HRa or People analytics as the combination of the simplest definitions of HRa [5, 6] and [7]: HR analytics is a data-driven framework for the application of analytical logic for HR management (HRM) function as the practice of data-driven HR decision-making.

The main objective of this study is to develop an evidence-based model for HRa consisting of the measurable factors that drive SMEs' performance in a way to have a better grasp on how to digitalize themselves. Two additional objectives are to identify the main challenges and consider the implications and recommendations of HRa implementation in SMEs.

This is precisely the basis for the starting hypothesis: H_1: Investing in HR analytics can increase organizational performance and accelerate the digital transformation of SMEs.

The approach taken by large enterprises is based on the premise that HR must be part of an AI platform. Organizations are required to apply an integrated approach that combines technology and skilled manpower to implement human resource analytical solutions for better results [11]. When dealing with SMEs, additional attention should be paid to avoid the uncritical replication of large firm ideas and ideals [2] and in that direction, the following hypothesis was put forward H_2: Uncritical replication of HRa implementation in large companies is counterproductive for SMEs.

2 Model of HRa Implementation in SMEs

The complex process of HRa implementation is multidimensional and involves different skill sets, methodologies, actors and infrastructures [7] No examples of HRa implementation models of SMEs can be found in the existing literature (peer-reviewed articles). It should be pointed out that the published results of the research indicate that the promises and predictions attributed to HRa are not being fulfilled [7]. "Despite the claimed importance of HR analytics, research investigating the performance impact of HR analytics on organizational performance remains underdeveloped" [8]. Also, the existing HRa literature remains underdeveloped in understanding whether and how such promises have been realized [9]. Some of the authors have stated that "one of the reasons could well be a failure to deliver added value" [3]. Others conclude that there is "very limited high-quality scientific evidence-based research on this topic." Finally, it should be concluded, "that this very limited research, as well as the promotional nature of the literature, can be partially explained by the relative newness of the field of HRa" [6].

To overcome this gap, we defined one solution: the HR People Analytic Model named HRPA (Fig. 1). We recognize HRa as a strategic management tool and the practice of data-driven HR decision-making that involves multiple stakeholders. This model includes external and internal factors, which should ensure the legitimacy of HRa. The basis of the HRPA model is the HRa process, which is carried out in 7 steps: goals setting - selection of metrics and the implementation team, determining the source of the data, data management, modelling–analytics, evaluating–implementation and reporting–dashboards. It is surrounded by external and internal factors related to the most important stakeholders of SMEs. In the middle are the organizational factors: organizational culture and skills.

3   Results

The HRPA model is based on the successful implementation of HRa, following the stages in the development of HR analytics [1], starting from the lowest level of HRa, descriptive analytics (HR 1.0), ending with Cognitive Analytics, which represents the highest level of HRa (HR 5.0). It is recommended to start introducing HR in SMEs sequentially, from HR 1.0, by implementing one of the HRM metrics as a pilot study.

For the model implementation, an ROI-based focus approach on defined metrics is proposed. When choosing the metrics for the pilot implementation, we suggest two options: according to strategically defined priorities, trying to apply to get the highest impact work done or choosing the activity and corresponding metrics that are the best supported by data (evidence-based).

The model was tested on real data from the company Tekijanka d.o.o. The employee retention metrics that provide insight into retention rate and likelihood to stay with the company were selected for testing. These metrics follow the organizational goal of keeping productive and talented workers and reducing turnover (primarily highly qualified). HR 1.0 was applied (descriptive analytics) using Tableau in the first phase; clustering, as a supervised machine learning technique in the HR 2.0 phase - diagnostic HRa; and log-linear regression as a supervised machine learning technique in the HR 3.0 phase - predictive HRa. The obtained results confirmed the initial hypotheses.

After the successful implementation of descriptive HRa, one should move on to diagnostic and predictive HRa on the same HRM function, leaving Prescriptive and Cognitive analytics for the final stage of implementation.

EXTERNAL FACTORS

STAKEHOLDERS	Government, competitors and business partners	Legitimacy
IMPLEMENTERS	External (outsourcing) companies	Vendors
CLOUD	ICT infrastructure	AI
	Data infrastructure	Confidential cloud

ORGANIZATIONAL FACTORS

ORGANIZATIONAL CULTURE – STRATEGIC DOCUMENTS – POLICES – PRIVACY – DATA PRIVACY – ETHIC	Process 1. Goals setting - Selection of metrics (based on defined strategic goals) 2. Selection of the implementation team (internal, external, mix) 3. Determining the source of the data 4. Data management (cleaning and other - special attention directed to BIAS) 5. Modeling - Analytics 6. Evaluating - Implementation 7. Reporting - Dashboards	Organizational skills – Strategic planning – Goal setting – Problem-solving – Teamwork – Training – Communication – Conflict management – "Soft skills"
DATA WAREHOUSE / DATA LAKE	ICT infrastructure	IT skills
	Data infrastructure	Data engineering
	Analytic	Data science
STAKEHOLDERS	Employees, owners	Management skills
IMPLEMENTERS	Team for the implementation	+ Domain knowledge + PM skills

DATA QUALITY

INTERNAL FACTORS

Fig. 1. HRPA - HR People Analytic Model

4 Conclusions

Literature review confirms the noted research gap: No examples of HRa implementation models of SMEs can be found in the existing literature (peer-reviewed articles). To achieve the effects expected from HRa, the paper proposes the standardization of the HRa implementation process in SMEs using the HRPA model (HR People Analytic Model). The proposed solution is based on the ROI-based focus approach in HRa implementation. The model foresees the successive implementation of HRa, starting from HR 1.0 (descriptive analytics) on the selected metrics to achieve the targets defined by the company's strategic documents. This enables organization insights into the set targets to support decision-makers.

The focus of future research will be on the detailed development, testing and improvement of the HRPA model. Towards the implementation of prescriptive and cognitive HRa, using big data analytics, reinforcement learning, AI and robotics towards the automation of routine HRa processes.

References

1. OECD, The Digital Transformation of SMEs, OECD Studies on SMEs and Entrepreneurship, OECD Publishing, Paris (2021). https://doi.org/10.1787/bdb9256a-en
2. Harney, B., Gilman, M., Mayson, S., Raby, S.: Advancing understanding of HRM in small and medium-sized enterprises (SMEs): critical questions and prospects. Int. J. Hum. Resource Manage. 33(16), 3175–3196 (2022). https://doi.org/10.1080/09585192.2022.21093752022
3. Ohlert, C., Giering, O., Kirchner, S.: Who is leading the digital transformation? Understanding the adoption of digital technologies in Germany. New Technol. Work Employ. 37(3), 445–468 (2022). https://doi.org/10.1111/ntwe.12244
4. Vukmirovic, D., Bolbotinovic, Z., Comic, T., Stanojevic, D.: HR analytics: Serbian perspective. In: 1st Serbian International Conference on Applied Artificial Intelligence (SICAAI), Kragujevac, Serbia (2022). http://www.aai2022.kg.ac.rs/aai-2022-papers/
5. Tomar, S., Gaur, M.: HR analytics in business: role, opportunities, and challenges of using it. J. Xi'an Univ. Archit. Technol. 12(7), 1299–1306 (2020). ISSN No: 1006–7930
6. Bhattacharyya, D.K.: HR Analytics: Understanding Theories and Applications. SAGE Publications, New Delhi (2017)
7. Belizón, M.J., Kieran, S.: Human resources analytics: a legitimacy process. Hum. Resour. Manag. J. 32, 603–630 (2022). https://doi.org/10.1111/1748-8583.12417
8. McCartney, S., Fu, N.: Bridging the gap: why, how and when HR analytics can impact organizational performance. Manage. Decis. 60(13), 25–47 (2022). Emerald Publishing Limited. https://doi.org/10.1108/MD-12-2020-1581
9. McCartney, S., Fu, N.: Promise versus reality: a systematic review of the ongoing debates in people analytics. J. Organ. Effectiveness: People Perform. 9(2), 281–311 (2022). Emerald Publishing Limited. https://doi.org/10.1108/JOEPP-01-2021-0013
10. Álvarez-Gutiérrez, F.J., Stone, D.L., Castaño, A.M., García-Izquierdo, A.L.: Human resources analytics: a systematic review from a sustainable management approach. Revista de Psicología del Trabajo y de las Organizaciones 38(3), 129–147 (2022). https://doi.org/10.5093/jwop2022a18
11. Opatha, H.H.D.P.J.: HR analytics: a literature review and new conceptual model. Int. J. Sci. Res. Publ. 10(06), 130–141 (2020). https://doi.org/10.29322/IJSRP.10.06.2020.p10217

Comparison of Data-Driven and Physics-Informed Neural Networks for Surrogate Modelling of the Huxley Muscle Model

Bogdan Milićević[1,2(✉)], Miloš Ivanović[2,3], Boban Stojanović[2,3],
Miljan Milošević[2,4,5], Vladimir Simić[2,4], Miloš Kojić[2,6,7], and Nenad Filipović[1,2]

[1] Faculty of Engineering, University of Kragujevac, Sestre Janjić 6, 34000 Kragujevac, Serbia
bogdan.milicevic@uni.kg.ac.rs, fica@kg.ac.rs
[2] Bioengineering Research and Development Center (BioIRC), Prvoslava Stojanovića 6, 34000 Kragujevac, Serbia
{mivanovic,miljan.m,vsimic}@kg.ac.rs,
boban.stojanovic@pmf.kg.ac.rs
[3] Faculty of Science, University of Kragujevac, Radoja Domanovića 12, 34000 Kragujevac, Serbia
[4] Institute for Information Technologies, University of Kragujevac, Jovana Cvijića bb, 34000 Kragujevac, Serbia
[5] Belgrade Metropolitan University, Tadeuša Košćuška 63, 11000 Belgrade, Serbia
[6] Serbian Academy of Sciences and Arts, Knez Mihailova 35, 11000 Belgrade, Serbia
[7] The Department of Nanomedicine, Houston Methodist Research Institute, 6670 Bertner Avenue, R7 117, Houston, TX 77030, USA

Abstract. Biophysical muscle models based on sliding filament and cross-bridge theory are called Huxley-type muscle models. The method of characteristics is typically used to solve Huxley's muscle contraction equation, which describes the distribution of attached myosin heads to the actin-binding sites, called cross-bridges. Once this equation is solved, we can determine the generated force and the stiffness of the muscle fibers, which can then be used at the macro level during finite element analysis. In our paper, we present alternative approaches to finding an approximate solution of Huxley's muscle contraction equation using neural networks. In one approach, we collect the data from simulations and train multilayer perceptrons to predict probabilities of cross-bridge formation based on the available actin site positions, time, activation, current and previous stretch. In another approach, besides using the data, we also inform the neural network with Huxley's equation, thus improving the generalization of the neural network's predictions.

Keywords: Huxley muscle model · physics-informed neural networks · numerical solving of partial differential equations · multi-scale modeling

© The Author(s), under exclusive license to Springer Nature Switzerland AG 2024
N. Filipović (Ed.): AAI 2023, LNNS 999, pp. 33–37, 2024.
https://doi.org/10.1007/978-3-031-60840-7_5

1 Introduction

Physics-informed neural networks (PINNs) are trained to handle supervised learning tasks while respecting any given physical principle described by general nonlinear partial differential equations [1]. These neural networks represent a brand-new family of data-efficient approximators for universal functions that easily encode any underlying physical laws as prior knowledge [1]. With PINN, a key innovation is the addition of a residual network that encodes the governing physics equations, and uses the output from a deep learning network, called a surrogate, to compute a residual value [2]. The neural network is trained to reduce the differential equation residual along with the residuals of initial and boundary conditions. PINNs use automated differentiation to compute differential operators on graphs.

The fundamental PINN formulation does not require labeled data, it results from other simulations, or experimental data. For PINNs, only the residual function calculation is necessary. It is also possible and sometimes required to provide simulation or experimental data for the network to be trained in a supervised way, particularly for inverse problems. The experimental or simulation data can also be used when boundary conditions or an Equation of State are missing to close a system of equations. After a PINN is trained, its inference can be used in scientific computing to replace conventional numerical solvers [2]. PINNs are a gridless technique because any point in the domain can be used as input without the need to define a mesh. Additionally, without having to be retrained, the trained PINN network can be used to predict the results on simulation grids with various resolutions [2]. Time-dependent issues can also benefit from the use of PINNs. Since time can be modeled as any other variable, it is possible to predict the output at a given moment without having to account for earlier time steps. For these reasons, unlike many conventional computational techniques, the computational cost does not scale with the number of grid points. PINN has been employed for predicting the solutions for the Burgers' equation, the Navier–Stokes equation, and the Schrodinger equation [3]. In this study, we solved Huxley's muscle equation, using PINN, to acquire the distribution of attached myosin heads to the actin-binding sites.

2 Methods

Huxley thought about the movements of the filaments within muscle and the likelihood of myosin heads connecting with actin filaments inside sarcomeres to form bridges (cross-bridges) [4]. The $n(x, t)$ function describes the rate of connections between myosin heads and actin filaments, as a function of the position of the nearest available actin-binding site relative to the equilibrium position of myosin head x:

$$\frac{\partial n(x, t)}{\partial t} - v \frac{\partial n(x, t)}{\partial t} = [1 - n(x, t)]f(x, a) - n(x, t)g(x), \forall x \in \Omega \qquad (1)$$

where $f(x,a)$ and $g(x)$ represent the attachment and detachment rates of cross-bridges respectively, v is the velocity of filaments sliding, calculated using current and previous stretch, and a is muscle activation given as a function of time. The partial differential Eq. (1) can be solved using the method of characteristics with the initial condition $n(x,0)$

$= 0$. Once the $n(x, t)$ values are acquired we can calculate force F within the muscle fiber and stiffness K using the equations:

$$F(t) = k \sum_{-\infty}^{\infty} n(x, t)xdx \text{ and } K(t) = k \sum_{-\infty}^{\infty} n(x, t)dx \tag{2}$$

where k is the stiffness of cross-bridges. Stress and stress derivative can be calculated as:

$$\sigma_m = F \frac{\sigma_{iso}}{F_{iso}} \text{ and } \frac{\partial \sigma_m}{\partial e} = \lambda L_0 K \frac{\sigma_{iso}}{F_{iso}}, \tag{3}$$

where F_{iso} is maximal force achieved during isometric conditions, σ_{iso} maximal stress achieved during isometric conditions, $L_0 the$ initial length of the sarcomere and λ is the stretch. Calculated stresses and stress derivatives can be further used at the macro-level during the finite element analysis. We used SciANN to implement PINN and integrate Eq. (1). This is a high-level artificial neural network API written in Python with Keras and TensorFlow backends. SciANN is designed to abstract the construction of neural networks for scientific computing and the solving and discovery of partial differential equations (PDEs) using physics-informed neural networks.

3 Results and Discussion

Using the SciANN framework we constructed a neural network with 8 layers, each containing 20 neurons with a hyperbolic tangent activation function. The network is trained by minimizing the difference between the actual and the predicted values and also by minimizing the residuals derived from Eq. (1) and its initial conditions. We used Adam optimizer with a learning rate of $5 \times 10{-}^5$ and batch size of 16384, during 7000 epochs. We also used the neural tangent kernel (NTK) method to get the adaptive weights, balancing the difference between the number of points used to minimize the residual of PDE, and the number of points used to minimize the residual of the initial condition. We also trained the ordinary multilayer perceptron (MLP) with the same architecture as PINN and we used the same data, but without providing the specificity of Huxley's muscle equation to the network.

Once the networks were trained, we integrated them into the finite element solver and used them at the micro-level instead of the method of characteristics. In Table 1, we show the correlation coefficients between original values, obtained in finite element simulations with the method of characteristics at the micro-level, and predicted values, obtained in simulations with the neural network at the micro-level. We presented acquired stresses and stress derivatives. These values were obtained in numerical experiments that were used to collect the data and train the neural networks. It can be seen that stresses and stress derivatives obtained with PINN are closer to the original values than the values obtained by MLP. In Table 2, numerical experiments that were not used in the training set are shown. It can be seen that PINN performed better in these experiments, which indicates that PINN generalizes better than the standard MLP.

Table 1. Correlation coefficients between original values, obtained by the method of characteristics, and predicted values, obtained by neural networks. Shown numerical experiments were used to acquire the data and train the neural networks.

Neural network:	PINN		MLP	
Identification number of numerical experiment	Correlation coefficient (stress)	Correlation coefficient (stress derivative)	Correlation coefficient (stress)	Correlation coefficient (stress derivative)
1	0.9929	0.9943	0.9852	0.9872
2	0.9860	0.9860	0.7782	0.8925
3	0.9972	0.9958	0.9902	0.9959
4	0.9343	0.9584	0.1286	0.8354
5	0.9817	0.9909	0.9964	0.9956
6	0.9962	0.9893	0.0884	0.8561
7	0.9978	0.9855	0.1559	0.7753
Average value:	**0.9837**	**0.9857**	0.5890	0.9054
Standard deviation:	**0.0209**	**0.0117**	0.4088	0.0823

Table 2. Correlation coefficients between original values, obtained by the method of characteristics, and predicted values, obtained by neural networks. Shown numerical experiments were used to test the neural networks.

Neural network:	PINN		MLP	
Identification number of numerical experiment	Correlation coefficient (stress)	Correlation coefficient (stress derivative)	Correlation coefficient (stress)	Correlation coefficient (stress derivative)
8	0.8861	0.9592	0.6968	0.8795
9	0.9704	0.9811	0.1854	0.7707
Average value:	**0.9283**	**0.9702**	0.4411	0.8251
Standard deviation:	**0.0421**	**0.0109**	0.2557	0.0544

4   Conclusions

In our article, we presented alternative methods to find approximate solutions of Huxley's muscle contraction equation using neural networks. We collected data from simulations and trained multilayer perceptron to predict cross-bridge formation probabilities. In addition to using the data, we also informed the neural network by calculating the residual of the Huxley equation, which resulted in an improvement of the neural network's ability to generalize predictions.

Acknowledgement. This research was supported by the European Union's Horizon 2020 research and innovation programme under grant agreement No 952603 (http://sgabu.eu/). This article reflects only the author's view. The Commission is not responsible for any use that may be made of the information it contains. Research was also supported by the SILICOFCM project that has received funding from the European Union's Horizon 2020 research and innovation programme under grant agreement No 777204. This article reflects only the authors' views. The European Commission is not responsible for any use that may be made of the information the article contains. The research was also funded by the Ministry of Education, Science and Technological Development of the Republic of Serbia, contract numbers [451-03-68/2022-14/200107 (Faculty of Engineering, University of Kragujevac) and 451-03-68/2022-14/200378 (Institute for Information Technologies Kragujevac, University of Kragujevac)].

References

1. Raissi, M., Perdikaris, P., Karniadakis, G.E.: Physics Informed Deep Learning (Part I): Data-Driven Solutions of Nonlinear Partial Differential Equations. New York City, NY (2017a). arXiv preprint arXiv:1711.10561
2. Markidis, S.: The old and the new: can physics-informed deep-learning replace traditional linear solvers? Front. Big Data **4**,(2021). https://doi.org/10.3389/fdata.2021.669097
3. Raissi, M., Perdikaris, P., Karniadakis, G.E.: Physics-informed neural networks: a deep learning framework for solving forward and inverse problems involving nonlinear partial differential equations. J. Comput. Phys. **378**, 686–707 (2019)
4. Williams, W.O.: Huxley's model of muscle contraction with compliance. J. Elast. **105**, 365–380 (2011)

Different Dataset Preprocessing Methods for Tree Detection Based on UAV Images and YOLOv8 Network

Milan Grujev[1][✉] ⓘ, Aleksandar Milosavljevic[1] ⓘ, Aleksandra Stojnev Ilic[1] ⓘ, Milos Ilic[2] ⓘ, and Petar Spalevic[2] ⓘ

[1] Faculty of Electronic Engineering, University of Nis, Aleksandra Medvedeva 14, 18000 Nis, Serbia
milan.grujev@elfak.rs
[2] Faculty of Technical Sciences, University of Pristina, Knjaza Milosa 7, 38220 Kosovska Mitrovica, Serbia

Abstract. The problem that we are trying to solve is related to determining the best approach for preparing a dataset to train a neural network for the needs of tree detection on agricultural areas. Namely, in order to analyze the condition of perennial orchards, a dataset was created by photographing agricultural areas with a drone and it consists of unprocessed, large-format images showing plantations of different ages and different planting structures. After the process of annotating the trees on the captured images, different models of automatic image preprocessing and augmentation were applied to prepare the dataset for the further training process. In order to compare different approaches of input dataset preparation, the training process was performed in the same way using the YOLOv8 neural network. The evaluation process of the obtained results has been performed by comparing metrics such as precision, recall, mAP@0.5, mAP@0.5–0.95, and F1score, for different versions of input datasets.

Keywords: YOLOv8 · tree detection · image augmentation · image processing

1 Introduction

The recent developments in the artificial intelligence domain have led to the proliferation of the studies targeting its application in various areas of industry, and daily human life and work. Precision agriculture relies on the application of modern software solutions in daily activities on agricultural plantations [1]. Such solutions contribute both in terms of reporting and forecasting methodology, as well as in the speed of processing data from the field and making conclusions about further agrotechnical measures that need to be implemented.

One of the ways of collecting information about the condition of agricultural crops is to record plantations with drones [2]. If perennial fruit plantations are observed, a large number of parameters, such as diversity in fruit growth and leaf color, can be obtained with this approach. If it is about large plantations, visiting the plantation to mark dried

© The Author(s), under exclusive license to Springer Nature Switzerland AG 2024
N. Filipović (Ed.): AAI 2023, LNNS 999, pp. 38–45, 2024.
https://doi.org/10.1007/978-3-031-60840-7_6

or poorly growing trees requires a lot of time [3]. Precisely for this reason, the main goal of this research is the preparation of the input set of images in order to obtain the most accurate detection of trees.

More precisely, for the needs of the research, a dataset of aerial photography of perennial plantations of different ages was created. In this way, the images that make up the raw dataset were obtained. As part of the research, we deal with the processing of such a set of data to create a set of images that will be most suitable for use in the process of neural network training and later use of the trained model for the need of tree detection.

The paper is organized as follows. The Sect. 2 presents an overview of the used literature that deals with a similar topic. Section 3 represents the methodology of the experimental part of the research. Section 4 presents obtained results and gives comparison of different approaches. Section 5 presents main conclusions and ideas for future research. The last section gives a references list.

2 Literature Review

Authors in [4] presented a system for tree detection and segmentation that recognizes four different types of trees. The authors presented two neural network models for different tasks. For tree detection, the authors used a modified YOLO model with a prediction grid. This model uses an RGB image of $256 \times 256 \times 3$ as input which is later split using a 5×5 grid. Every cell is responsible for recognizing one object. Every object is predicted as one or more bounding boxes with confidence factor and one-hot vector that represents the class of object, in this example, the tree species. The authors set the confidence factor value to 0.8. Dataset used for the training process was split into training/validation/test sets in a ratio 60/5/35. Second part of this system is the tree segmentation task. The segmentation model is based on a modified SegNet model. The input size of the segmentation model is also a $256 \times 256 \times 3$ image. The proposed model consists of a set of hierarchical convolutional modules to reduce size and gain more channels. The module contains three to five convolutional layers with batch normalization, and at the end of the module, it has one pooling layer and one activation layer. These compressed images present an input for a set of up-sampling modules. Up sampling modules contain one up-sampling layer and followed by several convolutional layers, and the last is the activation layer. Evaluation metrics that authors used for the proposed model is F1 score and classification accuracy. Authors achieved classification level of 97.5% and value 0.89 as highest F1 score value.

In one of the papers authors published a study about bayberry trees detection method based on improved YOLO algorithm [5]. This paper proposes an optimal YOLOv4 method for bayberry tree detection based on UAV images. The authors used a YOLOv4 architecture with CSPDarknet53 as a backbone network. This network consists of the convolutional and residual blocks. Authors proposed the Leaky ReLu activation function. In this study, the K-Means clustering method was directly embedded into the model to automatically obtain anchors, filter and delete unreasonable object boxes. Then combined with DIoU NMS method to remove the duplicate boxes. The authors created a dataset of 611 images selected from 3108 original images using pix4DMapper to process UAV data. The UAV image has a single size of 5472 × 3648 pixels. According to the image resolution and pixel size of the dataset, the data was clipped by 20% area overlap. The input image has a size of 1368 × 912 pixels. The training set accounts for 85% of the total dataset, and the validation set accounts for 80% of the training set. The test set accounts for 15% of the total dataset. The authors tried several samples from original images on the proposed model and used F1 score, precision and recall as evaluation metrics. For the F1 score, the highest value achieved was 98.16%. The highest achieved value for precision and recall was 97.78% and 98.16% respectively.

Similar model was proposed in [2]. In their paper, they used the YOLOv4 architecture model for coconut tree detection from UAV images. The authors used standard YOLOv4 model with CSPDarknet53 as backbone network. This model is fine-tuned on specific datasets to save training cost and time. Authors proposed smaller convolution kernels and stride for convolutional layers. For the purpose of training the model, authors have drone captured images with size 3000 × 4000. Datasets were manually cropped to 544 × 544 around the sections of the image where the trees were present. The authors used F1 score, precision, and recall for model evaluation. The F1 score achieved by the model is 93.79% with a recall value of 92%. For precision, the highest achieved value is 96.65%.

3   Methodology

For the purposes of this research, a set of images was created using the Phantom DJI Pro 4 drone. Photographing was performed on perennial plantations of one, four and eight years old. The number of trees per hectare was 666, 1250 and 1000 respectively, which shows that the structure and spacing of the trees is different on different plantations. The recording resulted in 62 large-sized images (approximately 5000 × 3000 px).

Further preparation of the input image set was performed using the Roboflow computer vision platform. Preprocessing of initial images resulted in six versions of the dataset by adjusting parameters such as image size, the number of tiles. Additional steps included image augmentation such as exposure. Exposure is applied just on the training dataset images. Detailed data on the method of image preparation process, as well as the number of images obtained after this process is shown in Table 1.

Created dataset is used in the training process of neural networks. For this purpose YOLOv8 neural network was used. YOLO is an object detection algorithm using a deep convolutional neural network. The initial release of YOLO was published in 2015. Several versions have been proposed since 2015, each building on and improving its predecessor. The latest version is YOLOv8 from 2023. The YOLO algorithm takes an image as input and then uses a simple deep convolutional network to detect the object. The architecture of the CNN model that forms the backbone of initial YOLO release consists of 24 convolutional layers, 4 max pooling layers and 2 fully-connected layers. The input is a fixed-size 448 × 448 image. This CNN model uses convolutional layers with 1 × 1 and 3 × 3 kernels.

Hidden layers use ReLu function for activation, except the final layer, which uses a linear activation function. Batch normalization and dropout are included as additional techniques for overfitting prevention [6].

YOLOv8 was released in January 2023 by Ultralytics, the company that developed YOLOv5. YOLOv8 provided five scaled versions: YOLOv8n (nano), YOLOv8s (small),

Table 1. Different versions of created dataset.

Version	Image size [pixels]	Tile [num]	Augmentation	Training set	Validation set	Test Set
V1	640 × 640	1 × 1	/	44	12	6
V2	640 × 640	4 × 4	/	704	192	96
V3	640 × 640	2 × 2	/	176	48	24
V4	640 × 640	2 × 2	Exposure 20%	516	48	24
V5	640 × 640	3 × 5	/	660	180	90
V6	640 × 640	3 × 5	/	660	180	90

YOLOv8m (medium), YOLOv8l (large) and YOLOv8x (extra large). YOLOv8 supports multiple vision tasks such as object detection, segmentation, pose estimation, tracking, and classification [7]. YOLOv8 uses a similar backbone as YOLOv5 with some changes on the CSPLayer, now called the C2f module. The C2f module (cross-stage partial bottleneck with two convolutions) combines high-level features with contextual information to improve detection accuracy. YOLOv8 uses an anchor-free model with a decoupled head to independently process objectness, classification, and regression tasks. This design allows each branch to focus on its task and improves the model's overall accuracy. In the output layer of YOLOv8, they used the sigmoid function as the activation function for the objectness score, representing the probability that the bounding box contains an object. It uses the softmax function for the class probabilities, representing the objects' probabilities belonging to each possible class. YOLOv8 can be run from the command line interface (CLI), or it can also be installed as a PIP package. In addition, it comes with multiple integrations for labeling, training, and deploying.

4 Results and Discussion

After creating different versions of the input dataset, training and evaluation of the YOLOv8 neural network was performed. The training process for each of the created versions of the input images was done in 50 epochs. For the purposes of validation and testing of neural networks trained in this way, the same images were always used. Table 2 shows the obtained evaluation results.

After a close comparison of the obtained evaluation parameters, we can conclude that version v4 gives the best results. Additional metrics for version v4 are shown on Fig. 1.

The test dataset included images that were not previously used in the training and validation process of the obtained model. The annotated images used for testing are shown in Fig. 2, while Fig. 3 shows the result of tree detection using the trained v4 model.

On Fig. 3, next to the labels, we have a confidence parameter that tells how sure the model is that it detected the tree. The research conducted included the creation of six models based on the same images. The difference between the created models is

Table 2. Evaluation results for different datasets.

Ver.	Precision	Recall	mAP@0.5	mAP@0.5–0.95	F1 score
V1	0.9398	0.8717	0.9217	0.4967	0.90
V2	0.8840	0.7351	0.8317	0.4497	0.80
V3	0.9217	0.8588	0.9216	0.5054	0.89
V4	0.9538	0.8502	0.9261	0.5236	0.90
V5	0.8596	0.7402	0.8254	0.4494	0.80
V6	0.8468	0.7304	0.8042	0.4146	0.78

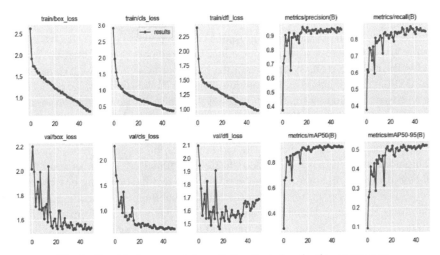

Fig. 1. Graphical representation of all collected evaluation parameters

reflected in the way of processing the input data set. Data sets created in this way were used in the process of training the model with the same algorithm. This kind of approach enabled an adequate comparison of the obtained results. At the same time, the use of the same evaluation methods provided the appropriate parameters, on the basis of which it is possible to show the differences between the trained models.

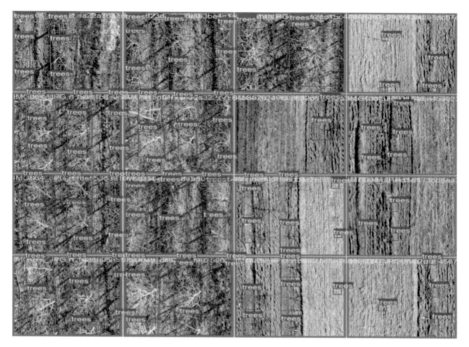

Fig. 2. Ground truth images

From the presented figures, it can be concluded that the trees were well detected with a high degree of confidence. The obtained results show that this approach is of great importance when it comes to the detection of trees in agricultural areas. This approach can provide much faster and more accurate detection of trees compared to traditional methods. This is especially true when it comes to large agricultural areas as well as areas that are not easy to reach. It is precisely for this reason that it is possible to apply this approach also on areas that represent an example for the sake of the forest.

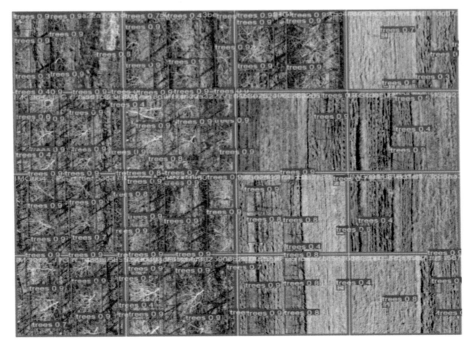

Fig. 3. Prediction images

5   Conclusion

Object detection methods could provide significant support in the precision agriculture process. The results of the conducted research showed that the selection of image pre-processing parameters, and augmentations parameters has a huge impact on the final prediction results. Future research will be focused on the segmentation of tree branches based on detected trees, so as to provide a mechanism for distinguishing between good and bad trees, based on their appearance. This approach can provide farmers and agricultural experts with information about the state of the trees, all with the aim of planning further production. In addition, based on a timely assessment of the condition of the plantations, the expected yields can be planned as well as the further processing of the products themselves.

References

1. Guang, C., Shang, Y.: Transformer for tree counting in areal images. Remote Sens. **14**, 1–16 (2022)
2. Kshitij, N.N., et al.: Modified YOLOv4 for real-time coconut trees detection from an unmanned aerial vehicle. Research Square, pp. 1–19 (2021)
3. Youliang, C., et al.: An object detection method for bayberry trees based on an improved YOLO algorithm. Int. J. Digit. Earth **16**(1), 781–805 (2023)
4. Ochoa, K.S., Guo, Z.: A framework for the management of agricultural resources with automated aerial imagery detection. Comput. Electron. Agric. **162**, 53–69 (2019)

5. Youliang, C., Hanli, X., Xiangjun, Z., Peng, G., Zhigang, X., Xiaobin, H.: An object detection method for bayberry trees based on an improved YOLO algorithm. Int. J. Digit. Earth **16**(1), 781–805 (2023)

6. Redmon, J., Divvala, S., Girshick, R., Farhadi, A.: You only look once: unified, real-time object detection. In: Proceedings of the IEEE Conference on Computer Vision and Pattern Recognition, pp. 779–788 (2016)

7. Terven, J.R., Cordova-Esparza, D.M.: A comprehensive review of YOLO: from YOLOv1 and beyond. arXiv 2023. arXiv preprint arXiv:2304.00501 (2023)

Data-Driven Digital Twins of Renewable Energy Grids

Danica N. Prodanovic[1]([⊠]), Nikola N. Andrijevic[1], Bosko R. Lakovic[2], and Boban S. Stojanovic[1,2]

[1] Faculty of Science, University of Kragujevac, Radoja Domanovica 12, Kragujevac, Serbia
{danica.prodanovic,nikola.andrijevic,
boban.stojanovic}@pmf.kg.ac.rs
[2] Vodena, Kralja Milana IV 19B/5, Kragujevac, Serbia
{bosko.lakovic,boban.stojanovic}@vodena.rs

Abstract. This conference paper presents the concept of data-driven digital twins for renewable energy grids, with a focus on the application of artificial intelligence methods for modeling photovoltaic systems. The proposed data-driven digital twin system utilizes automated machine learning algorithms to generate accurate models of solar collectors, enabling real-time monitoring and forecasting of energy production. A use case of a solar power plant in a company from Čačak, Serbia is presented, demonstrating the potential of this technology to improve energy management and increase the efficiency of renewable energy sources. Results obtained from the use case demonstrate the effectiveness of this approach in optimizing energy management, reducing energy consumption, and improving the reliability of renewable energy systems.

Keywords: artificial intelligence · renewable energy · modeling · forecasting · digital twins

1 Introduction

As the world continues to shift towards renewable energy sources, the integration of renewable energy grids into existing power systems has become an increasingly complex challenge. In order to effectively manage these complex systems, digital twin technology has emerged as a promising solution. Digital twins are virtual replicas of physical assets, systems, or processes that are used to simulate and optimize their real-world behavior.

In recent years, significant effort has been made in research of digital twins of renewable energy grids and application of artificial intelligence in modeling renewable energy assets. H. Xu et al. [1] present a comprehensive review of data-driven digital twins for renewable energy systems, discussing the key components of such systems and the potential for their application in the industry. A comprehensive review of the application of artificial intelligence in renewable energy systems, discussing the various approaches and techniques used to optimize energy management is given in Kim et al. [2]. Dara and Panda [3] discuss the use of artificial intelligence in renewable energy forecasting, highlighting the benefits of using machine learning algorithms to improve the accuracy of renewable energy predictions.

The goal of this paper is to provide an understanding of data-driven digital twins for renewable energy grids, and to demonstrate their potential to transform the way we manage and optimize renewable energy systems. In the paper, we discuss the key components of a digital twin system, including data acquisition, modeling, and simulation. We also examine the benefits and challenges of implementing digital twins in renewable energy grids and present a case study that highlights their potential impact.

2 Methodology

The introduction of renewable energy sources (RES) in the grid has posed several challenges to energy producers and consumers, such as intermittency effect, "duck curve", the growing complexity of stimuli regulations and the calculation of energy consumption. All these challenges change the role of distribution network operators from energy transmitters into "orchestrators" of a large number of prosumers, which differ in size, type, and patterns of consumption and production. Orchestration and optimization of such complex energy systems using traditional methods is no longer feasible. These challenges require the application of novel approaches based on the synergy of IoT and AI.

2.1 Concept of the Digital Twin

Existing digital twins have mostly relied on physically based models that are established on the fundamental physical principles that govern the behavior of the modeled grid assets. Physical models are highly accurate but have many drawbacks, which limit their utilization. They heavily depend on the vast number of parameters that need to be detailed and perfectly reflect the modeled system to give accurate predictions. Furthermore, physically based models cannot take every single aspect of the system into consideration. Aspects such as the external state and performance degradation are not strictly governed by laws of physics and therefore cannot be precisely simulated.

To solve this problem, we have developed a comprehensive data-driven energy management platform that completely automates finding an optimal pattern in energy consumption and production in case of facilities with RES and energy storage capabilities. Based on the data acquired during the grid exploitation, it automatically creates the most adequate predictive models of the internal energy production and loads, which enable simulation of the data chain for any hypothetical operation plan. The results obtained from the simulations, along with all other grid features and external factors, are subjected to an optimization process in order to find the energy management pattern that will result in the most economical usage of electricity under given conditions.

To create and maintain models of all the assets within the grid, we employ Black Fox [4] cloud service for automated generation of optimized machine learning models, based on Deep Neural Networks, Random Forest and XGBoost. Black Fox performs genetic algorithm (GA) optimization of all elements of the machine learning model with the aim of generating a model that best describes input data. For the sake of shortness, within this paper we will present the results of the automated ML modeling of solar power plants only, but similar principles can be applied to all other assets within a grid.

2.2   Automatically Created Model of Solar Power Plant

Input Data

The quality of a machine learning model depends on the accuracy and availability of the data that goes into the training process. Through our extensive research on the available weather data providers, we have determined that Solcast service offers the best solution for accurate and available weather data that will be used for simulating the behavior of solar panels in renewable energy grids.

The Solcast service provides accurate and reliable weather data for any location in the world. Solcast's weather data is based on reliable sources, such as satellite images and ground-based observations, and is updated frequently to reflect the latest weather patterns. The weather data includes various weather variables such as temperature, humidity, cloud coverage, wind speed, and solar radiation. These variables are crucial for accurately predicting the energy output of solar panels in a renewable energy grid. The data is collected in real-time and has hourly granulation, providing us with detailed insights into the behavior of the solar panels. The data is also presented in a clear and user-friendly format, making it easy for users to interpret and apply to their specific needs. In addition to their high-quality weather data, Solcast offers a range of tools and services, including forecasting models and APIs, to further enhance the accuracy and accessibility of their data.

For collecting forecast data locally, we created a service that periodically downloads weather data from Solcast and saves the data into the database specialized for time series data. The specialized database offers easy and quick download, search, and visualization of data. The weather data collection service is developed to automatically recover from errors and is resilient to failures for up to 7 days.

ANN Model

For the purposes of creating an artificial neural network able to model the production of electricity from solar panels, historical meteorological data collected using the Solcast service and the measured production of a solar power plant in a company from Cacak, Serbia, were used. The following table holds information about historical meteorological and time data used for the development of the artificial neural network (Table 1).

Table 1. Input parameters

Input parameter	Description
DayOfYearSin	Sine function representation of day of year
HourOfDaySin	Sine function representation of hour of day
DayOfYearCos	Cosine function representation of day of year
HourOfDayCos	Cosine function representation of hour of day
Tamb	Air temperature
Cloud opacity	Cloud coverage
DHI	Direct Horizontal Irradiance
DNI	Direct Normal Irradiance
EBH	Horizontal component of DNI
GHI	Global Horizontal Irradiance

The artificial neural network for modeling electricity production uses the previously described parameter as an input, and as an output it provides an estimate of the system's electricity production under given conditions. All available data were used for training, divided into training, test and validation sets. The network was trained in 3000 epochs with an enabled early stopping mechanism. For a loss function the mean squared error was used, with the Adam training algorithm. The architecture of the network was obtained through optimization of parameters and hyperparameters using software for automated generation of the most adequate predictive model for any given data set, Blackfox.

3   Results and Validation

To accurately determine the extent to which the network can produce satisfactory predictions, a thorough investigation was performed. It analyzed the impact of varying training set sizes on the quality of predictions. Data related to the recorded production of the PONS solar power plant were used. The network was first trained on a small training set of only 100 data records (little more than 4 days) and tested on data related to the next seven days. Then the initial training set was extended for records related to the next thirty days, and the next seven days were again selected for the test set. This process is repeated as long as there is available production data.

For validation of automatically created ANN models we have used an equivalent physically based model. Physically based models are established on the fundamental physical principles that govern the behavior of solar panels. These models use mathematical equations to describe the physical processes that occur within a solar panel, such as the absorption of sunlight and the transport of charge carriers. To compare the forecasts obtained by ANN we have used a physically based model from PVLIB library, one of the largest and most comprehensive collections of such models.

The prediction quality of all trained networks was quantified by RMSE metric, which is depicted on the next figure (Fig. 1).

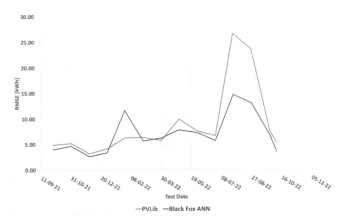

Fig. 1. The influence of the size of the training set on the performance of the network.

Upon the inspection, it is evident that the artificial neural network outperforms the physically based PVlib library, even when utilizing smaller training set sizes. The figure also shows an uncharacteristic error spike of the network, but it is currently unknown what caused this abnormality.

4   Conclusions

In this paper we have presented a concept of data-driven digital twins of renewable energy grids and proven the concept of modeling solar power plants using automated machine learning. The results obtained using the automatically created and optimized ANN have shown significantly better performance over the referent PVlib physically based model. During the long period of time the ANN model has shown robustness in situations when the external factors altered the power plant performance. In contrast to the physically based model, the ANN model was able to adapt to changes such as snow coverage, upgrade of the power plant, and solar panel soiling, making it more effective for the application in real-world settings.

Acknowledgement. This paper is funded through the EIT's HEI Initiative SMART-2M project, supported by EIT RawMaterials, funded by the European Union.

References

1. Xu, H., et al.: Data-driven digital twins for renewable energy systems: a review. Appl. Energy **284**, 116081 (2021)
2. Kim, S., et al.: Artificial intelligence for renewable energy systems: review. Appl. Energy **257**, 114052 (2020)
3. Dara, S.S., Panda, S.: A review on artificial intelligence applications in renewable energy forecasting. Renew. Sustain. Energy Rev. **117**, 109507 (2020)
4. Blackfox. www.blackfox.ai

AI Anomaly Detection for Smart Manufacturing

Bojana Bajic[1,2(✉)], Milovan Medojevic[2], Milos Jovicic[2], and Aleksandar Rikalovic[1,2]

[1] Department of Industrial Engineering and Management, Faculty of Technical Sciences, University of Novi Sad, Novi Sad, Serbia
{bojana.bajic,a.rikalovic}@uns.ac.rs, {bojana.bajic, aleksandar.rikalovic}@ivi.ac.rs
[2] Institute for Artificial Intelligence Research and Developments of Serbia, 21000 Novi Sad, Serbia
{milovan.medojevic,milos.jovicic}@ivi.ac.rs

Abstract. AI-based anomaly detection represents a significant research field with myriad application possibilities in smart manufacturing that involves using machine learning (ML) algorithms to identify abnormal patterns in the data collected from sensors and other sources. While substantial progress has been made in this field in recent years, some gaps in the literature still need to be addressed. One key challenge is the lack of labeled data for training ML algorithms, especially for rare anomalies. Moreover, unsupervised anomaly detection is also an obstacle to smart manufacturing implementation, as it can be prone to false positives or missing some anomalies. Additionally, the imbalanced dataset, where the collected data samples are predominantly from one specific class, results in no specific information about the anomalies that must be detected. Addressing these gaps, the present research provides a comparative analysis of three unsupervised ML techniques, namely: One-class Support Vector Machine – OCSVM, Isolation Forest – IF, and Local and Outlier Factor – LOF for product anomaly detection. The ML models were developed using an imbalanced dataset collected in the manufacturing company from the process industry. Finally, the results show the OCSVM technique provides the best performance regarding accuracy, precision, recall, and F1 score.

Keywords: unsupervised learning · artificial intelligence (AI) · Industry 5.0 · smart manufacturing · big data analytics (BDA) · machine learning (ML)

1 Introduction

Nowadays, anomaly detection represents one of the most popular applications of AI in industrial environments. It involves using machine learning algorithms to identify abnormal patterns in the data collected from sensors and other sources in the manufacturing process [1, 2]. These abnormal patterns could indicate equipment failure, product defects, or other issues that could affect the quality and efficiency of the manufacturing process. Overall, AI-based anomaly detection is a valuable tool for manufacturers looking to improve the quality, efficiency, and safety of their operations [1]. By detecting anomalies early and taking corrective actions, manufacturers can reduce downtime, minimize waste, and improve overall productivity.

© The Author(s), under exclusive license to Springer Nature Switzerland AG 2024
N. Filipović (Ed.): AAI 2023, LNNS 999, pp. 52–56, 2024.
https://doi.org/10.1007/978-3-031-60840-7_8

However, while AI-based anomaly detection has made significant progress in recent years, there are still implementation challenges addressed in the literature [3, 4] that researchers are working to address representing the major gap. The two key gaps in anomaly detection research include:

- *imbalanced data:* In present days, achieving a balance between the number of non-fault and fault data samples is challenging in manufacturing companies [5]. This is especially true for process industries that collect mostly non-fault data. However, that kind of dataset cannot be analyzed using supervised methods such as linear regression, support vector machines, and artificial neural networks since they require a balance between faulty and non-faulty data [6]. Additionally, the non-fault data often consists of numerous similar data values that do not offer varied information about the manufacturing system's condition [7], and
- *unsupervised anomaly detection:* While supervised learning algorithms can be effective at detecting anomalies in labeled data, unsupervised anomaly detection is still a challenge. Unsupervised learning algorithms rely on clustering and other techniques to identify anomalies, but these methods can be prone to false positives or miss some anomalies [8].

Thus, the present research aims to fill these gaps by conducting a comparative analysis of the three state-of-the-art unsupervised ML techniques, namely: One-class Support Vector Machine (OCSVM), Isolation Forest (IF), and Local Outlier Factor (LOF) for products anomaly detection. The unsupervised ML models were developed and tested based on real manufacturing data collected from the process industry company. Finally, the collected manufacturing data represent the imbalanced industrial dataset where most of the collected process data samples belong to one specific class without information on the specific anomalies that need to be detected.

2 Methodology

The developed methodology was inspired by data mining. Notably, the methodology development was informed by the practical field experience of the research team in the implementation of smart manufacturing systems in industry. The developed methodology consists of three phases (Fig. 1), namely:

Phase 1: Dataset optimization empowered by Edge computing – focuses on optimizing the dataset, which is carried out through three distinct steps. The first step involves defining the smart manufacturing problem to gain a clear understanding of the processes and activities involved. Next, the parameters are identified based on expert knowledge and experience, and finally, the data is preprocessed using edge computing. The primary goal of this phase is to improve production resilience and streamline big data optimization by rationalizing power and processing resources using edge computing technology. This is achieved by reducing the size of the big data to a smaller, precisely selected dataset that still contains the essential information from the original dataset. This selected dataset serves as the basis for establishing the anomaly detection model [9];

Phase 2: Anomaly detection models development – consists of two distinct steps. The first step involves dividing the dataset into 80:20 training and testing subsets [10]. The next refers to the selection of the unsupervised ML models that are developed using state-of-the-art anomaly detection methods, namely: OCSVM, IF, and LOF;

Phase 3: Anomaly detection models testing – refers to the validation of the unsupervised ML anomaly detection models, and their results are compared. In the final step, the performance of the anomaly detection model is evaluated by measuring the following metrics [11, 12]:

- *accuracy*, that measures the proportion of correct predictions out of the total number of predictions;
- *precision*, that measures proportion of correctly identified positive samples among all predicted positive samples;
- *recall*, that measures the proportion of true positives out of the total number of actual positive samples;
- *F1 score*, that harmonic mean of precision and recall and is a good measure of overall model performance, especially when the classes are imbalanced [13].

These metrics are among the conventional evaluation methods [14] used in classification and clustering problems.

Fig. 1. Anomaly detection methodology for smart manufacturing.

3   Dataset Optimization Empowered by Edge Computing

The dataset used in this study was obtained from a real manufacturing environment form the process industry. The original dataset includes 15 significant process parameters as inputs selected based on expert knowledge, where in total, 29,403,000 data units were collected and grouped into 6,534 separate.csv files. The collected dataset was optimized based on methodology proposed in [9]. This resulted in the optimization of 29,403,000 data units from the 15 most important process parameters. The optimization was performed using edge computing technology, achieving a 99.73% optimization rate without loosing significant information contained in big data. The resulting optimized dataset consisted of a single.csv file containing 3,802 data samples from the 12 most important process parameters, resulting in a final dataset of 45,624 data units.

4   Unsupervised ML Models Development and Testing

The optimized dataset is divided into a training set and a testing set in an 80:20 ratio, which is considered optimal for developing anomaly detection models [3]. 80% of the dataset is used for training the model, while 20% is used for testing. Using process parameters collected from a selected company, we developed predictive models for product anomaly detection. It is important to note that the collected dataset was highly imbalanced, as the case company primarily produced high-quality products. To address the issue of an imbalanced dataset, we employed three methods for anomaly detection, including OCSVM, IF, and LOF, and compared obtained results (Table 1). We kept all other experimental parameters unchanged to ensure comparability.

Table 1. Evaluation performances of the anomaly detection using unsupervised ML techniques.

Results	OCSVM	IF	LOF
Accuracy	96.32%	89.87%	95.00%
Precision	100%	100%	100%
Recall	96.32%	89.87%	95.00%
F1 score	98.12%	94.66%	97.44%

The obtained results of the accuracy, precision, recall, and F1 score measurements (Table 1) suggest that OCSVM outperformed all other ML models, although there is minimal difference in the results. Notably, performance metrics such as accuracy, precision, recall, and F1 score each measure a different aspect of an algorithm's performance. In practical applications, the F1 score is widely used as the most reliable measurement, especially when dealing with imbalanced datasets in industrial conditions. Given the performance evaluation results and the significant weight placed on the F1 score, it is proven that OCSVM model can be deployed in the manufacturing environment and assess its performance in real-time.

5 Conclusions

In the present research, we proposed AI-based methodology for anomaly detection inspired by data mining. The proposed methodology consists of three phases: 1) *Dataset optimization empowered by Edge computing*; 2) *Anomaly detection models development*; and 3) *Anomaly detection models testing*. In the following paragraphs, we summarize the main contributions of the present research.

ML model development based on imbalanced industrial dataset: The relevant literature addressed the implementation challenges of smart manufacturing [3, 4] based on ML technologies. The main reason for this gap is the poor quality of datasets collected in manufacturing environments, which can significantly aggravate the application of ML methods. To address this gap, this research demonstrates that by using imbalanced industrial data, it is possible to develop ML models with high reliability.

Comparison of the state-of-the-art unsupervised ML models for anomaly detection: During the development of the model, three different state-of-the-art anomaly detection methods for imbalanced dataset were used, namely: OCSVM, IF, and

LOF. Notably, OCSVM model outperformed all other ML models regarding accuracy, precision, recall and F1 score. Noticeably, the obtained results indicate that OCSVM has a strong potential for solving specific engineering problems by analyzing process data to detect and diagnose anomalies in manufacturing systems.

References

1. Abdelrahman, O., Keikhosrokiani, P.: Assembly line anomaly detection and root cause analysis using machine learning. IEEE Access **8**, 189661–189672 (2020)
2. Cheng, C., Chen, P.: Phase I analysis of nonlinear profiles using anomaly detection techniques. Appl. Sci. **13**, 2147 (2023)
3. Bajic, B., Rikalovic, A., Suzic, N., Piuri, V.: Industry 4.0 implementation challenges and opportunities: a managerial perspective. IEEE Syst. J. **15**(1), 546–559 (2021)
4. Rikalovic, A., Suzic, N., Bajic, B., Piuri, V.: Industry 4.0 implementation challenges and opportunities: a technological perspective. IEEE Syst. J. **16**(2), 2797–2810 (2022)
5. Dai, H.N., Wang, H., Xu, G., Wan, J., Imran, M.: Big data analytics for manufacturing Internet of Things: opportunities, challenges and enabling technologies. Enterp. Inf. Syst. **14**, 1–25 (2019)
6. Bajic, B., Suzic, N., Simeunovic, N., Moraca, S., Rikalovic, A.: Real-time data analytics edge computing application for Industry 4.0: the Mahalanobis-Taguchi approach. Int. J. Ind. Eng. Manag. **11**(3), 146–156 (2020)
7. Klikowski, J., Woźniak, M.: Employing one-class SVM classifier ensemble for imbalanced data stream classification. In: International Conference on Computational Science, pp. 117–127 (2020)
8. Bahri, M., Salutari, F., Putina, A., Sozio, M.: AutoML: state of the art with a focus on anomaly detection, challenges, and research directions. Int. J. Data Sci. Anal. **14**(2), 113–126 (2022)
9. Bajic, B., Suzic, N., Moraca, S., Stefanovi, M.: Edge computing data optimization for smart quality management: Industry 5.0 perspective. Sustainability **15**(7), 6032 (2023)
10. Gholamy, A., Kreinovich, V., Kosheleva, O.: Why 70/30 or 80/20 relation between training and testing sets: a pedagogical explanation (2018)
11. Watanabe, T., Kono, I., Onozuka, H.: Anomaly detection methods in turning based on motor data analysis. Procedia Manuf. **48**(2019), 882–893 (2020)
12. Wang, J., Gao, S., Tang, Z., Tan, D., Cao, B., Fan, J.: A context-aware recommendation system for improving manufacturing process modeling. J. Intell. Manuf. **34**(3), 1347–1368 (2023)
13. Hackeling, G.: Mastering Machine Learning with Scikit-Learn, 1st ed. Packt Publishing, Birmingham (2014)
14. Abdullahi, A., Samsudin, N.A., Ibrahim, M.R., Aripin, M.S., Khalid, S.K.A., Othman, Z.A.: Towards IR4.0 implementation in e-manufacturing: artificial intelligence application in steel plate fault detection. Indones. J. Electr. Eng. Comput. Sci. **20**(1), 430–436 (2020)

Drawbacks of Programming Dataflow Architectures and Methods to Overcome Them

Nenad Korolija[✉]

School of Electrical Engineering, University of Belgrade, Bulevar Kralja Aleksandra 73, Belgrade, Serbia
nenadko@etf.rs

Abstract. Compared to the control-flow paradigm dataflow paradigm is proven to be superior for certain classes of high-performance computing algorithms, such as AI. However, this comes with the cost. Many problems should be solved in order to efficiently utilize dataflow hardware. Some of them are programming dataflow architectures and debugging issues that arise during this process, relatively long simulation time, finding suitable dataflow hardware, and difficulties with the debugging hardware problems that might occur. This paper analyses existing solutions for accelerating high-performance computing algorithms along with problems that each architecture introduces, as well as potential solutions to the aforementioned drawbacks of utilizing the dataflow paradigm for accelerating algorithms. The analysis is performed on both hardware and software levels. Results indicate that dataflow hardware has the potential to be efficiently used along with control-flow hardware using AI. The best performance is achieved when combining control-flow and dataflow hardware on the same chip die.

Keywords: High-Performance Computing · Control-flow · Dataflow

1 Introduction

Control-flow algorithms are based on principles defined by von Neumann [1]. The control-flow paradigm has existed almost since the first computers were introduced. At that time, it seemed most suitable for processing relatively simple algorithms that computers were used for.

The control-flow computer program consists of consecutive instructions, that are often executed independently using the control-flow type of computer architecture. Executing each instruction consists of a few phases: fetching the instruction, fetching operands, executing the arithmetical operation in the case of arithmetic instruction, storing the result, and checking whether an interrupt has occurred. Fetching an instruction utilizes an internal bus and register file to load the instruction from the memory into the instruction register. Fetching operands utilize the same parts of architectures, except that the resulting operands may be loaded into another register. Storing the result again utilizes an internal bus, and often a register file. This limits the algorithm parallelism.

Supporting the execution of any instruction defined by the control-flow architecture along with caching mechanisms results in utilizing around a billion transistors on

© The Author(s), under exclusive license to Springer Nature Switzerland AG 2024
N. Filipović (Ed.): AAI 2023, LNNS 999, pp. 57–70, 2024.
https://doi.org/10.1007/978-3-031-60840-7_9

a chip. However, the number of instructions that can be executed per clock cycle is approximately the same as the number of multi-core processor (CPU) cores.

Demands by computer graphics lead to the development of so-called manycore processors, consisting of thousands of cores. Later, they were utilized in executing high-performance computing algorithms, enabling algorithm execution speedups up to a couple of orders of magnitude. However, manycore processors suffer from the same problems as any other architecture based on the control-flow paradigm.

The arithmetic logic unit (ALU) in a control-flow processor is responsible for performing arithmetic and logic functions. Most of ALUs work with 64-bit operands. If we compare the number of transistors utilized for performing an operation, we can notice that there could be a million adders implemented using a billion transistors [2]. Therefore, it can be concluded that the improvement of processors is far from reaching the limit of the technology.

In contrast to the control-flow paradigm, where each instruction execution is strictly controlled, the dataflow paradigm is based on data flowing through the hardware configured for a particular algorithm [3]. This brings multiple benefits, as well as constraints. The main characteristics of dataflow hardware and their consequences are:

- All instructions implemented in hardware can run simultaneously,
- Data representing the result of executing an instruction travels to portions of hardware that execute instructions that need this result. Therefore, the internal bus is not a bottleneck of dataflow hardware,
- It is suitable only for certain classes of high-performance algorithms, including AI algorithms,
- Only highly scalable portion of algorithms are suitable for the dataflow hardware,
- Control-flow hardware is usually needed beside dataflow hardware,
- The hardware has to be configured before starting the execution,
- The number of instructions that can be executed is limited by the number of transistors at the chip,
- Utilizing dataflow hardware for multiple algorithms requires re-configurable hardware,
- The frequency of dataflow hardware is an order of magnitude lower than those of control-flow processors.

Numerous algorithms can be accelerated using the dataflow paradigm [4–9]. Algorithms [4–8] are accelerated using the same framework. The main prerequisite for the dataflow hardware to be able to accelerate certain control-flow algorithms is that it has many repetitions in executing a set of instructions. In some cases, these instructions include a relatively high amount of processing of the same data, while in others they access data from arrays, which is faster using the dedicated hardware compared to the general, control-flow hardware.

From the programmer's point of view, programming control-flow hardware requires way fewer skills compared to programming dataflow hardware. Control-flow hardware benefits from decades of improvements by a large programmers community. On the other side, dataflow hardware was predominantly programmed using specialized hardware languages.

The recent development of dataflow programming languages enables programming dataflow hardware as if it were control-flow hardware. There are frameworks for automatically translating statements from high-level language into the VHDL program and further into the hardware configurations which are available on the web, allowing a programmer to develop a dataflow algorithm without possessing the hardware [10].

Work has been done on automating the translation of control-flow programs into the dataflow representation [11, 12], so that the programmer, or engineer, needs only to govern the process, instead of defining the dataflow hardware configuration thoroughly.

The following sections define the research problem, present existing solutions, the proposed solution, the analysis, and conclusions, as proposed in the paper by Milutinovic et al. [13].

2 Problem Statement

This work has multiple sometimes opposing goals. A general goal may be defined as accelerating high-performance computing algorithms at reasonable costs.

The most important aspect of high-performance computing algorithms is their execution time. The execution should be as fast as possible. High-performance computing algorithms are often highly scalable and the highest processing speeds are often achieved using as many processing elements as the algorithm scalability allows.

The importance of the cost of the computing system cannot be overstated. Specialized hardware is often so expensive that it is utilized only in highly profitable projects. For example, high-frequency trading requires fast computer reasoning and response. The company owning the fastest hardware has an advantage over its competitors. On the other hand, we are all witnessing using cheaper hardware for coin mining, as the profit cannot always justify the hardware cost.

In many programming projects, the salaries of engineers are almost directly proportional to the costs of the projects. The amount of work needed to finish the project can be
measured using many techniques. When it comes to programming dataflow hardware,
it requires considerably more effort than programming control-flow hardware [14]. The
constraints in programming dataflow hardware could often be automatically taken care
of. For example, the amount of resources can be taken into account by the framework for
translating a dataflow kernel written in a dedicated dataflow language into the dataflow
hardware configuration. However, there are more sophisticated actions that should be
performed in order to deliver optimal performance of the hardware. For example, Algorithm 1 for processing a matrix calculates each matrix element value based on the value
of the element to the left. In this case, a simple control-flow program can evaluate matrix
elements row by row, as it is more efficient due to temporal and spatial locality. Access-
ing a value that has been already accessed in a previous for-loop iteration is expected to
result in a cache hit. In opposition to this scenario, a Dataflow kernel that would process
elements in this order would stall in a considerable amount of the total matrix processing
time. This is because the dataflow pipeline would have to stall until the result is produced
on the output before processing the next element that needs this result. This is depicted
in Fig. 1. Algorithm 1 presents a sample code to demonstrate how stalls occur. Rows are
being streamed to the kernel one by one. We can consider a single row. While row[j−1],
which corresponds to A[i, j−1], is being processed in stages, the following input to the
kernel, row[j] could be provided only once all stages are processed. Therefore, only a
single stage of the kernel can be active at any given moment.

```
for (int i = 0; i < nRows; i++){

    for(int j = 0; j < nColumns; j++){

        float vFactor = (j > 0 ? 0.1 : 0 );

        if(j)

            A[i][j] = A[i][j-1] * vFactor + A[i][j];

    }

}
```

Algorithm 1: Processing a matrix by rows

Algorithm 2 depicts MaxJ language implementation of the dataflow kernel executing
the equivalent code to the control-flow algorithm given in Algorithm 1. One can note
that the offset function requires a previous value of a row to be provided in order to
calculate the current one. This clearly shows that before the rowOut is calculated and
sent as the output, the following element can not be computed.

```
public class StreamKernel extends Kernel {

public StreamKernel (KernelParameters parameters) {

    super(parameters);

    HWVar row = io.scalarInput("row" ,hwFloat(8, 24));

    ...

    // 32 bit counter with module 100:

    HWVar cntFactor = control.count.simpleCounter(32, 100);


    HWVar vFactor > 0 ? 0.1 : 0;

    HWVar rowOut = cntFactor ?

        stream.offset(row, -1) * vFactor + row : row;

    ...

    io.output("rowOut", rowOut, hwFloat(8, 24));

    ...

}

}
```

Algorithm 2: Dataflow implementation with dependent iterations

If function *func* consists of e.g. 1000 stages and the kernel is the only one running on the dataflow hardware, the utilization of the dataflow hardware will be approximately 0.1% of its capabilities. This condition can be loosened if the result from the previous iteration is not needed at the beginning of the following iteration, but the problem still exists. In order to efficiently use dataflow hardware, preferably, there should be no stalls at all. The frequencies of dataflow hardware are often an order of magnitude lower compared to the control-flow processors. Therefore, if a dataflow hardware executes a single instruction at any given moment, it is expected that the total execution of the algorithm will be even more than 10 times slower using the dataflow hardware because of the frequencies of both types of hardware, as well as the communication cost between the control-flow and the dataflow hardware.

The recent trend in increasing the cost of electrical energy along with work on reducing pollution sets the power efficiently as one of the most important aspects of high-performance computing architectures. Lowering power consumption also reduces the necessity for cooling processors.

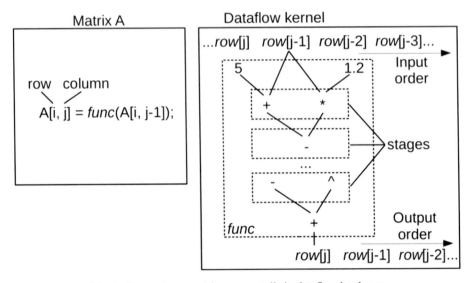

Fig. 1. Processing a matrix causes stalls in dataflow hardware.

The size of the computer system is important for many reasons. High-performance computing architectures can occupy a building. Keeping it in working condition represents a considerable cost. Another reason why the size of the computer system matters is that the distance between processing elements affects the communication speed [15].

Last, but not least, we are all aware that any technology has its peaks, but also its declines. Programming projects are more and more often sold as services, where a user pays for the usage, not for the software itself. If the technology is outdated, it is often more reasonable to build the project from scratch, than to apply the old code to the new hardware. Therefore, it is important to choose the technology that has the potential to be utilized for a reasonably long time period.

3   Existing Solutions

Control-flow architectures were traditionally exploited in high performance computing architectures. They were capable of executing any algorithm. The acceleration of a scalable algorithm execution is achieved by parallelizing the execution on multiple processors. Various improvements have been introduced in decades, including new cache architectures [16].

Clusters consist of racks with nodes, each containing a certain number of control-flow processors. They are specialized in scalable algorithms. Nodes often offer more memory compared to those found in desktop computers, but also more processing power.

With the rising capabilities of computer graphics cards, so-called manycore processors found their place in parallel processing. The advantage over conventional cluster processing arose from the fact that cores are close to each other, so that the communication between them is faster than communication between cluster nodes.

In recent history, we have witnessed the growing significance of cloud processing. Relatively big corporations are leading in shaping cloud environments. If we consider only cloud solutions for high-speed processing, they offer the possibility to rent certain amounts of hardware resources in terms of the number of processors with certain characteristics, or in terms of virtual capacities, but they also offer cloud services, i.e. software as a service (SAAS).

Technology advancements in recent decades enabled the efficient use of dataflow architectures in accelerating certain classes of high-performance computing algorithms. Application-specific integrated circuit (ASIC) chips are customized for particular use cases. They are known to be smarter, faster, and more reliable in specific domains, e.g., networking, but they are also used in computing demanding applications, e.g., bitcoin mining. In contrast to ASIC, field-programmable gate arrays (FPGA) are re-configurable chips that can be utilized for multiple algorithms. Architectures based on FPGA often appear in the form of dataflow cards connected over peripheral component interconnect express (PCIe) bus, but there are also dataflow hardware with different organization, including ring structure and dataflow cards connected directly to the network.

In recent years, researchers have been focusing not only on increasing the number of instructions that can be executed in a second, but also on accelerating algorithms that operate on the ever-increasing amount of data [17, 18]. Also, hybrid control-flow and dataflow processors have been introduced [19] that combine both control-flow and dataflow hardware on the same chip die, often accompanied by additional hardware accelerators. Researchers often suggest incorporating multicore and dataflow [20, 21] hardware, and sometimes even manycore at the same chip [22–24]. The resulting architecture has faster communication between these three components compared to the conventional approach, where they would be connected over PCIe. Numerous scientific fields can benefit from these advancements [25–27]. Algorithms that can be accelerated using the dataflow paradigm in such a way that they imply frequent communication between control-flow and dataflow hardware can be divided based on the fact whether the total time spent on computation is higher than the time spent for sending data between these two types of hardware. If the dataflow hardware is waiting for the communication to finish before starting the processing of data, it is a clear signal that the improvement in the communication speed might reduce the total algorithm execution time. Other algorithms might experience the same execution speed as if they were run using the dataflow hardware connected to the control-flow hardware using the PCIe bus.

From the programmer's point of view, as mentioned in the introduction section, the control-flow paradigm benefits from decades of improvements by the community. On the opposite, dataflow hardware is harder to work with for many reasons:

- Engineers need to know not only programming, but also hardware constraints,
- There are fewer programming utilities,
- Development is usually performed on simulators that are slower than the hardware,
- Debugging is harder due to difficulties in extracting signals from chips,
- Scheduling for dataflow introduces new constraints [28], etc.

4 Proposed Solution

The dataflow hardware has better performance in uniform processing compared to control-flow hardware. However, the constraint is that dataflow hardware is usually not suitable for executing the whole high-performance computing algorithm, but only the part that includes a relatively high amount of repetition. Therefore, the control-flow hardware is needed for preparing the data for the dataflow hardware and handling the results.

The configuration of the FPGA can be achieved using the control-flow hardware. For example, Groq designs chips with dataflow accelerators based on FPGA with support for control-flow software to control dataflow hardware.

The distance between the control-flow and the dataflow hardware of a single computer system limits the parallelism due to the communication speed. This not only reduces the number of algorithms for which it is justifiable to utilize dataflow hardware, but also reduces the speed of configuring the dataflow hardware. Therefore, hybrid control-flow and dataflow processors are proposed to reduce the communication gap between these two types of hardware by using shared cache memory [29, 30], enabling fine algorithm granularity [31].

Regarding the software, the automation of translation from higher level language into the dataflow hardware configuration using frameworks brought several advantages compared to programming in VHDL:

- Eliminated necessity for a programmer to know any hardware description language,
- Eliminated necessity for a programmer to know hardware specifics,
- Considerably reduced programming time,
- Easier and faster debugging,
- Framework orchestrates the communication between the CPU and the dataflow hardware, while the programmer only initializes the communication by calling functions from included library.

Recently, work has been done on automating the translating process from a control-flow algorithm into the dataflow algorithm by detecting portions of source code (e.g. loops) that can be translated into dataflow kernels and empirically evaluating which portions should be the best to put on the dataflow. The prerequisite is that the programmer knows what kind of input data he expects to be working with, so that the system is optimized for such inputs. This process can be greatly improved by using AI to detect portions of source code that should be executed using the dataflow hardware.

Another aspect where AI could be of crucial importance is rearranging the computation in order to improve the utilization of the dataflow hardware. For example, if matrix elements are processed row by row, as in the example from Fig. 1, the control-flow source code could be modified so that the matrix is processed column by column. This way, while processing element $A[i, j]$, the kernel can start processing $A[i+1, j]$, then $A[i+2, j]$, avoiding stalls if the number of rows is greater than the number of stages. This is depicted in Fig. 2. While $A[i, j]$ is being calculated based on $A[i, j-1]$ in stages, one stage after another, the processing of elements of different rows can be done in other stages in parallel. This way, if the number of rows is higher or equal to the number of stages, once the A[i, j] is needed as the input to the dataflow kernel, it will already be calculated by the kernel. This way, stalls are avoided.

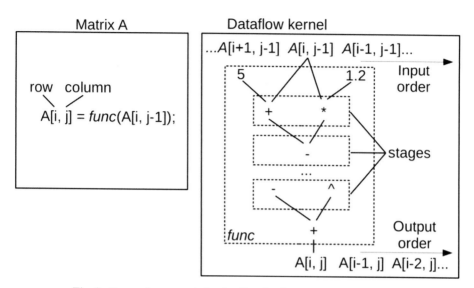

Fig. 2. Processing a matrix by dataflow hardware column by column.

Detecting the possibility to reorganize the computation might seem a bit fuzzy and relatively hard to achieve, but there are tools that can help. For a C++ source code from Algorithm 1, a dependency graph can be constructed, showing dependencies between constants and variables in the for loop. If switching positions of the inner and outer loop, as shown in Algorithm 3, eliminates or relaxes the dependency between consecutive iterations, then reorganizing the computation is expected to speed up the execution of the algorithm on dataflow hardware.

```
for(int j = 0; j < nColumns; j++){
  for (int i = 0; i < nRows; i++){
    float vFactor = (j > 0 ? 0.1 : 0 );
    if(j)
      A[i][j] = A[i][j-1] * vFactor + A[i][j];
  }
}
```

Algorithm 3: Processing a matrix by columns.

Both Figs. 1 and 2 include operations in the first two stages that can be optimized out. Instead of adding the number 5 to the matrix element, and then subtracting 1.2 times the same matrix element, one could solely add the number 5 to the −0.2 times the matrix element. Source codes with the same set of operations but different numbers can be automatically generated, serving as a synthetic benchmark for training AI networks to learn how to optimize them out. In order to avoid false positives, similar source codes that produce different outputs could also be automatically generated to train the AI network. Details of this optimization are out of the scope of this paper. The only purpose of the example is to depict how AI can be used to further optimize dataflow kernels.

5   Analysis

The proposed solution effectively solves problems that occur when using control-flow hardware for running high-performance computing algorithms that are suitable for dataflow hardware.

Example applications that are developed using the same framework can be found online in AppGallery project [32–34]. Sample applications include:

– Huxley Muscle Model,
– Correlation,
– Poisson Solver,
– High Speed Packet Capture,
– Brain Network,
– Linear Regression,
– Classification,
– Packet Pusher.

Some of these applications have source code available online, while others are not publicly available. Buttons CPU SRC are enabled for those applications that have CPU source available. Buttons DFE SRC are enabled for those applications that have dataflow kernels source available. Besides these, there are GIT buttons for github repositories, as well as TECH buttons for technical presentations. MaxJ language is a part of the dataflow framework for programming kernels. It is similar to Java programming language, but it also supports defining hardware variables. Hardware variables can be set to values from software variables, but the opposite is not allowed, as hardware variables represent the voltage level on wires, which does not necessarily have minimal (ground level) voltage or maximal voltage, but can be anywhere in between.

As already stated, this work has multiple sometimes opposing goals. This section focuses on analyzing various aspects of the proposed solution.

The hybrid control-flow and dataflow processor incorporates the dataflow hardware. As a result, it offers comparable speedup to dataflow hardware utilized with control-flow processors. Besides running high-performance computing algorithms that are known to have the possibility to be effectively accelerated using the dataflow hardware, the hybrid processor also enables accelerating fine-grained algorithms, where the execution of only a few lines of code could be efficiently accelerated using the dataflow hardware.

Currently, processors based on control-flow are produced much more often than the dataflow hardware cards for accelerating high-performance computing algorithms. As a result, the dataflow hardware is still more expensive, but massive production on hybrid processors is expected to lower the price of dataflow hardware, i. e. the price of processors should be comparable to those already available. However, a lot of work has to precede proposed hybrid processors.

As programming dataflow hardware requires considerably more effort than programming control-flow hardware, the automation of translation of high-performance computer algorithms already implemented for the control-flow into the dataflow is necessary, so that any application suitable for the dataflow hardware could be implemented relatively fast based on the control-flow version. Currently, only projects whose budget allows investing relatively high amounts into dataflow programmers generate algorithms for the dataflow hardware. Having in mind the cost of the programming for dataflow hardware, it is no surprise that the source code of algorithms already implemented for the dataflow hardware is usually not publicly available.

The dataflow paradigm is inherently parallel, which results in higher dataflow hardware processing speed measured in instructions per second compared to the control-flow hardware. With comparable electrical power of both types of hardware, the considerably faster processing leads to lower power consumption. With lower power consumption, the necessity for cooling is also reduced.

Packing more dense hardware for executing instructions reduces the size of the hardware needed for executing an algorithm. Further, the reduced distance between processing elements also speeds up the communication.

With all of the benefits of proposed hybrid processors, as well as the constant improvement and rising demand for AI applications, it is expected that the need for hybrid processors will grow over time. However, there is a concern about the reliability of the hybrid processor implementation, as the proper operation of the hardware depends on the operation of both types of hardware. As a result, it is expected that the lifespan will be minimal of the two hardware, but the frequency of counterfeits will be minimal as well [35, 36]. Additionally, the introduction of hybrid processors brings additional constraints due to the extra bus needed for the communication between the two types of hardware available on the same chip die [37]. This raises various questions including synchronizing timing for the two types of hardware, providing space on the chip, synchronizing accesses to the bus, etc. However, most of these problems are already solved in the existing control-flow hardware available today.

6   Conclusions

The focus of this work is on multiple aspects of using dataflow hardware for accelerating high-performance computing algorithms. By analyzing the von Neumann principles of control-flow hardware, it can be seen that it is not the most suitable architecture for algorithms with a high level of repeating patterns. The dataflow paradigm assumes that the data flows through the preconfigured hardware, which lowers the requirements for a portion of hardware necessary for executing a single instruction. However, dataflow hardware usually cannot execute an algorithm on its own, but depends on control-flow hardware. Naturally, the hybrid control-flow and dataflow processor is proposed for solving the gap in the communication between these two types of hardware.

Programming dataflow hardware usually requires more skills than programming control-flow hardware. The classical approach to programming dataflow hardware requires that a programmer must know both the specifics of the hardware and a hardware programming language. Nowadays, a framework can automatically translate a kernel programmed in a high-level programming language into the VHDL and further configure the reconfigurable hardware based on FPGA. Additionally, work has been done on automating the translation of portions of control-flow algorithms into the dataflow hardware, which enabled the automatic selection of portions of control-flow algorithms suitable for dataflow hardware.

The proposed approach effectively solves problems of accelerating algorithms, the cost of programming dataflow hardware, the power consumption of high-performance computing, as well as the size of the computer system. The cost of the computing system is expected to be comparable to the control-flow hardware, once the production level increases enough, which could be expected due to the potential of the proposed technology.

Acknowledgment. Author is partially supported by the School of Electrical Engineering, University of Belgrade, Serbia, the Institute of Physics Belgrade, contract no. 0801-1264/1, and the Ministry of Education, Science, and Technological Development of the Republic of Serbia.

References

1. Arikpo, I.I., Ogban, F.U., Eteng, I.E.: Von neumann architecture and modern computers. Global J. Math. Sci. **6**(2), 97–103 (2007)
2. Iannucci, R.A.: A critique of multiprocessing von Neumann style. ACM SIGARCH Comput. Archit. News **11**(3), 426–436 (1983)
3. Milutinović, V., Salom, J., Trifunović, N., Giorgi, R.: Guide to Dataflow Supercomputing, vol. 10, pp. 978–983. Springer, New York (2015). https://doi.org/10.1007/978-3-319-16229-4
4. Stojanovic, S., Bojic, D., Milutinovic, V.: Solving gross Pitaevskii equation using dataflow paradigm. Trans. Internet Res. **9**(2) (2013)
5. Kos, A., Rankovic, V., Tomazic, S.: Sorting networks on Maxeler dataflow supercomputing systems. Adv. Comput. **96**, 139–186 (2015). Amsterdam, Elsevier, Academic Press
6. Korolija, N., Djukic, T., Milutinovic, V., Filipovic, N.: Accelerating Lattice-Boltzman method using Maxeler DataFlow approach. Trans. Internet Res. **9**(2), 5–10 (2013)
7. Stanojevic, I., Senk, V., Milutinovic, V.: Application of Maxeler dataflow supercomputing to spherical code design. Trans. Internet Res. **9**(2), 1–4 (2013)
8. Bezanic, N., Popovic-Bozovic, J., Milutinovic, V., Popovic, I.: Implementation of the RSA algorithm on a DataFlow architecture. Trans. Internet Res. **9**(2), 11–16 (2013)
9. Ngom, A., Stojmenovic, I., Milutinovic, V.: STRIP-a strip-based neural-network growth algorithm for learning multiple-valued functions. IEEE Trans. Neural Networks **12**(2), 212–227 (2001)
10. Korolija, N., Zamuda, A.: On cloud-supported web-based integrated development environment for programming DataFlow architectures. In: Milutinovic, V., Kotlar, M. (eds.) Exploring the DataFlow Supercomputing Paradigm. CCN, pp. 41–51. Springer, Cham (2019). https://doi.org/10.1007/978-3-030-13803-5_2

11. Korolija, N., Popović, J., Cvetanović, M., Bojović, M.: Dataflow-based parallelization of control-flow algorithms. Adv. Comput. **104**, 73–124 (2017). Elsevier
12. Milutinovic, V., Salom, J., Veljovic, D., Korolija, N., Markovic, D., Petrovic, L.: Transforming applications from the control flow to the dataflow paradigm. In: Dataflow Supercomputing Essentials, pp. 107–129. Springer, Cham (2017). https://doi.org/10.1007/978-3-319-661 28-5_4
13. Milutinovic, V.: The best method for presentation of research results. IEEE TCCA Newsl. 1–6 (1996)
14. Popovic, J., Bojic, D., Korolija, N.: Analysis of task effort estimation accuracy based on use case point size. IET Software **9**(6), 166–173 (2015)
15. Trobec, R., et al.: Interconnection networks in petascale computer systems: a survey. ACM Comput. Surv. (CSUR) **49**(3), 1–24 (2016)
16. Milutinovic, V., Tomasevic, M., Markovi, B., Tremblay, M.: A new cache architecture concept: the split temporal/spatial cache. In: Proceedings of 8th Mediterranean Electrotechnical Conference on Industrial Applications in Power Systems, Computer Science and Telecommunications (MELECON 1996), vol. 2. IEEE (1996)
17. Flynn, M.J., et al.: Moving from petaflops to petadata. Commun. ACM **56**(5), 39–42 (2013)
18. Milutinović, V., Furht, B., Obradović, Z., Korolija, N.: Advances in high performance computing and related issues. Math. Probl. Eng. **2016**, 1–3 (2016)
19. Yazdanpanah, F., Alvarez-Martinez, C., Jimenez-Gonzalez, D., Etsion, Y.: Hybrid dataflow/von-Neumann architectures. IEEE Trans. Parallel Distrib. Syst. **25**(6), 1489–1509 (2013)
20. Popović, J., Jelisavčić, V., Korolija, N.: Hybrid supercomputing architectures for artificial intelligence: analysis of potentials. In: 1st Serbian International Conference on Applied Artificial Intelligence (SICAAI), Kragujevac, Serbia (2022)
21. Miladinović, D., Bojović, M., Jelisavčić, V., Korolija, N.: Hybrid manycore dataflow processor. In: Proceedings, IX International Conference IcETRAN, Novi Pazar, Serbia, 6–9 June 2022
22. Milutinović, V., et al.: The ultimate dataflow for ultimate supercomputers-on-a-chip, for scientific computing, geo physics, complex mathematics, and information processing. In: 10th Mediterranean Conference on Embedded Computing, pp. 1–6. IEEE, June 2021
23. Milutinović, V., et al.: The ultimate data flow for ultimate Super Computers-on-a-Chip. In: Handbook of Research on Methodologies and Applications of Supercomputing, pp. 312–318. IGI Global (2021)
24. Milutinović, V., Trifunović, N., Korolija, N., Popović, J., Bojić, D.: Accelerating program execution using hybrid control flow and dataflow architectures. In: 2017 25th Telecommunication Forum (TELFOR), pp. 1–4. IEEE, November 2017
25. Babović, Z., et al.: Research in computing-intensive simulations for nature-oriented civil-engineering and related scientific fields, using machine learning and big data: an overview of open problems. J. Big Data **10**(1), 1–21 (2023)
26. Babović, Z., et al.: Teaching computing for complex problems in civil engineering and geosciences using big data and machine learning: synergizing four different computing paradigms and four different management domains. J. Big Data **10**(1), 89 (2023)
27. Egharevba, L., Kumar, S., Amini, H., Adjouadi, M., Rishe, N.: Detecting and removing clouds affected regions from satellite images using deep learning. IPSI BGD Trans. Internet Res. **19**(2), 13–23 (2023)
28. Korolija, N., Bojić, D., Hurson, A.R., Milutinovic, V.: A runtime job scheduling algorithm for cluster architectures with dataflow accelerators. In: Advances in Computers, p. 126 (2022). Elsevier
29. Korolija, N., Milfeld, K.: Towards hybrid supercomputing architectures. J. Comput. Forensic Sci. **1**(1), 47–54 (2022)

30. Milutinovic, V., Markovic, B., Tomasevic, M., Tremblay, M.: The split temporal/spatial cache: a complexity analysis. Proc. SCIzzL **6**, 89–96 (1996)
31. Popović, M., Korolija, N., Štrbac-Savić, S.: Hybrid control-flow and dataflow processor: algorithm granularity analysis. In: Zbornik 29. konferencije YUINFO (2023)
32. Trifunovic, N., Milutinovic, V., Korolija, N., Gaydadjiev, G.: An AppGallery for dataflow computing. J. Big Data **3**(1), 1–30 (2016)
33. Trifunovic, N., Perovic, B., Trifunovic, P., Babovic, Z., Hurson, A.R.: A novel infrastructure for synergistic dataflow research, development, education, and deployment: the Maxeler AppGallery project. Adv. Comput. **106**, 167–213 (2017). Elsevier
34. Milutinovic, V., Salom, J., Veljovic, D., Korolija, N., Markovic, D., Petrovic, L.: Maxeler AppGallery Revisited. DataFlow Supercomputing Essentials: Research, Development and Education, pp. 3–18 (2017)
35. Huang, K., Liu, Y., Korolija, N., Carulli, J.M., Makris, Y.: Recycled IC detection based on statistical methods. IEEE Trans. Comput. Aided Des. Integr. Circuits Syst. **34**(6), 947–960 (2015)
36. Huang, K., Liu, Y., Korolija, N., Carulli, J.M., Makris, Y.: Statistical methods for detecting recycled electronics: from ICs to PCBs and beyond. IEEE Des. Test **41**, 15–22 (2023)
37. Milfeld, K., Korolija, N.: Towards hybrid supercomputing architectures. J. Comput. Forensic Sci. **1**(1), 47–54 (2022)

Forecasting River Water Levels Influenced by Hydropower Plant Daily Operations Using Artificial Neural Networks

Miloš Milašinović[1](✉), Dušan Marjanović[2], Dušan Prodanović[1],
and Nikola Milivojević[2]

[1] Faculty of Civil Engineering, University of Belgrade, Bulevar Kralja Aleksandra 73, 11000 Belgrade, Serbia
{mmilasinovic,dprodanovic}@grf.bg.ac.rs
[2] Jaroslav Černi Water Institute, Jaroslava Černog 80, 11226 Belgrade, Pinosava, Serbia
{dusan.marjanovic,nikola.milivojevic}@jcerni.rs

Abstract. Multipurpose water systems are used to deal with multiple objectives related to the usage of water for daily human activities. These activities are often conflicted which creates a challenging water management task. To provide reliable water resources management decision support tools for successful forecasting of hydraulic data (river flows and water levels) are essential. This research presents an approach for forecasting river water levels influenced by hydropower plant operations using artificial neural networks. This approach estimates hourly water level fluctuations at the control location using the water levels and hydropower plant discharge data as input. This tool can be used for fast assessment of different hydropower plant operation plans and help in choosing the optimal one. This water level forecasting procedure is applied and tested on the Iron Gate water system, placed on the Danube River, to deal with multiple objectives in water system management (hydropower production, flood protection, and inland navigation) and shows promising results.

Keywords: River water level · Forecasting · Artificial Neural Networks · Danube River · Iron Gate

1 Introduction

Climate change, energy transition and population growth put additional pressure in water resources management. This can lead to making multipurpose water systems too sensitive on different control operations, where inappropriate operation plans by one of the users can significantly reduce other objectives thus reducing the system overall performance. Therefore, operators in charge require reliable tools for fast assessment of the system operations and its influence on system's state. When river water systems are analyzed, water levels at different locations are often used to describe the system's state. Hence, decision support tools must be able to assess water levels affected by different operating scenarios.

N. Filipović (Ed.): AAI 2023, LNNS 999, pp. 71–75, 2024.
https://doi.org/10.1007/978-3-031-60840-7_10

Nowadays, forecasting of the river hydraulic data (river flows and water levels) is conducted using data driven techniques, such as artificial neural networks [1–4]. These artificial neural networks (ANN) applications show good results in forecasting hydraulic data, but particular focus is placed only on the rivers in natural conditions. To expand the ANN application range, river conditions affected by anthropogenic factors (such as multi-purpose water systems) have to be analyzed. This research focuses on forecasting of river water levels affected by hydropower plant operations. The idea is to analyze the ability of fast estimation of river water levels on a control station using hydropower plant discharge as an input data. The goal of this research is to provide reliable forecasting tool to support optimal hydropower plant (HPP) scheduling [5–8]. The proposed method is tested on the transboundary Iron Gate water system, placed on Danube River between Serbia and Romania.

2   Materials and Methods

2.1   Case Study

Transboundary Iron Gate water system is placed on Danube River between Serbia and Romania. Upstream effects of daily operations (scheduled discharge) have to be evaluated on a specific locations on a daily basis. One of the monitoring locations is Nera control station, placed 132 km upstream from the Iron Gate dam (Fig. 1.).

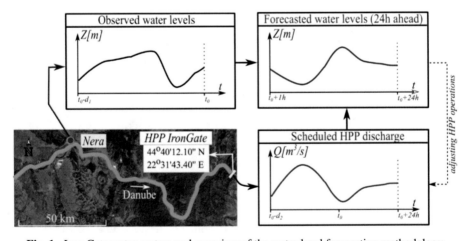

Fig. 1. Iron Gate water system and overview of the water level forecasting methodology.

2.2   Artificial Neural Network for Water Level Forecasting

For 24 h ahead water level forecasting at Nera station, with hourly timestep, recurrent neural network, in a form of Nonlinear AutoRegressive network with eXogenous inputs

(NARX) is used. HPP scheduled discharge is used as an exogenous input and previous water levels are used as feedback inputs, as presented in the following equation:

$$Z^t = f\left(Z^{t-1}, \ldots, Z^{t-d_1}, Q_{HPP}^{t-1}, \ldots, Q_{HPP}^{t-d_2}\right) \tag{1}$$

In this equation, predicted value of a water level Z^t at time t is regressed on given d_1 (feedback delay) past values of water level and d_2 (input delay) past values of an independent (exogeneous) input presented by HPP discharge Q_{HPP}. In this research, d_1 is set to 4h and d_2 is set to 24h.

ANN is created using two hidden layers consisting of 20 neurons each (Fig. 2) in MATLAB programming environment [9]. Number of hidden layers and number of neurons is arbitrary selected to demonstrate the methodology. Effects of different set of ANN parameters should be a analyzed in a separate research. To train the ANN, Levenberg-Marquardt backpropagation algorithm is used, along with hyperbolic tangent sigmoid transfer function and mean squared error (*MSE*) loss function.

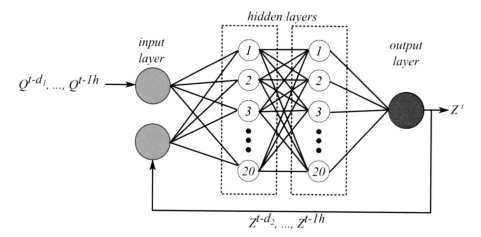

Fig. 2. ANN configuration.

3 Results and Discussion

HPP discharge and Nera water levels historical data for the period between June 4th 2017 and November 2nd 2019 are used as the training and validation sets. To test the ANN, data for period between November 3rd 2019 and November 19th 2019 is used. The proposed ANN (Eq. 1 and Fig. 2) is used to forecast daily water level timeseries for each day in training set, and the results are merged into one timeseries for the entire test period (Fig. 3). Results show that ANN can reproduce trends in water levels fluctuations well according to root mean square error values (*RMSE*). Daily *RMSE* values vary between 1.6 and 9.2 cm, while average *RMSE* value for the entire test period is 4.4 cm. Even though the proposed ANN performed well, for further improvement of the forecasting procedure, thorough analysis of the ANN parameters (input and feedback delays, ANN hyperparameters) has to be conducted along with enriching the training set with more recent data.

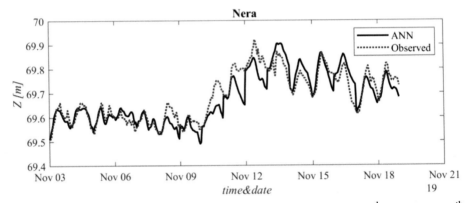

Fig. 3. Water levels at Nera hydrological station for the period November 3rd – November 19th 2019.

4   Conclusions

This research presents the potential of using ANNs to forecast river water levels at specific control locations affected by daily hydropower plant operations. Recurrent neural network has been utilized for 24 h ahead water level forecasting at Nera station, 132 km upstream from the Iron Gate dam. Hydropower plant discharge is used as the independent (exogeneous) input to ANN, along with past values of the water level. This research shows good results considering the *RMSE* indicator, but further investigations have to be conducted to improve forecasts.

Even though water levels across the river domain can be estimated using physically based (hydraulic) models, this approach requires a large set of data considering the boundary conditions and model parameters. This approach can produce bad results due to numerous uncertainties in the input data. When there is a necessity to assess water levels only at specific control points at the river section, and historical water level and hydropower plant discharge data is available, data driven techniques, such as artificial neural networks (proposed in this research), can be used instead. Finally, data driven techniques shouldn't be considered as a substitute for physically based models. Real world problems, such as complex river water system management, require hydraulic data forecasting on additional locations, where historical data for training data driven models is not available. In that case, physically based models and data driven methods have to be coupled into a hybrid model that can provide reliable forecasting on the entire river domain (data driven forecasting used as a boundary condition for hydraulic model). Proposing the framework for coupling these approaches for river water level forecasting will be the subject of forthcoming research.

References

1. Sung, J.Y., Lee, J., Chung, I.M., Heo, J.H.: Hourly water level forecasting at tributary affecteby main river condition. Water (Switzerland) **9**(9), 1–17 (2017)
2. Niedzielski, T., Miziński, B.: Real-time hydrograph modelling in the upper Nysa Kłodzka river basin (SW Poland): a two-model hydrologic ensemble prediction approach. Stoch. Environ. Res. Risk Assess. **31**(6), 1555–1576 (2017)
3. Kimura, N., Yoshinaga, I., Sekijima, K., Azechi, I., Baba, D.: Convolutional neural network coupled with a transfer-learning approach for time-series flood predictions. Water (Switzerland) **12**, 96 (2019)
4. Le, X.H., Ho, H.V., Lee, G., Jung, S.: Application of long short-term memory (LSTM) neural network for flood forecasting. Water (Switzerland) **11**(7), 1387 (2019)
5. Ćirović, V., Bogdanović, D., Bartoš-Divac, V., Stefanović, D., Milašinović, M.: Decision support system for Iron Gate hydropower system operations. In: Contemporary Water Management: Challenges and Directions, pp. 293–309 (2022)
6. Milašinović, M., Prodanović, D., Zindović, B., Stojanović, B., Milivojević, N.: Control theory-based data assimilation for hydraulic models as a decision support tool for hydropower systems: sequential, multi-metric tuning of the controllers. J. Hydroinformatics **23**(3), 500–516 (2021)
7. Milašinović, M., Prodanović, D., Stanić, M., Zindović, B., Stojanović, B., Milivojević, N.: Control theory-based data assimilation for open channel hydraulic models: tuning PID controllers using multi-objective optimization. J. Hydroinformatics **24**(4), 898–916 (2022)
8. Jaroslav Černi Water Institute & University of Belgrade - Faculty of Civil Engineering, Hydroinformatic system for Iron Gate hydropower plant (Published in Serbian), Belgrade, MA, 2019–2020
9. MathWorks Inc., Matlab Version 8.6. 0.267246 (R2018b). The MathWorks Inc Natick, MA (2018)

Enhancement of Outdated Vision Systems in SMEs with Artificial Intelligence Powered Solutions

Nemanja Pajić[1,2(✉)], Lazar Pavlović[2,3], Fatima Živić[1], and Jovana Aleksić[4]

[1] Faculty of Engineering, University of Kragujevac, Sestre Janjic 6, Kragujevac, Serbia
nemanjapajic2@gmail.com, zivic@kg.ac.rs
[2] ZF Serbia doo - ZF Friedrichshafen AG, 7. Nova, Pančevo, Serbia
[3] School of Electrical and Computer Engineering of Applied Studies, Belgrade, Serbia
[4] Faculty of Mechanical Engineering, Belgrade, Serbia

Abstract. This paper deals with the application of Artificial Intelligence (AI) based software solutions to the quality assurance in manufacturing of the printed circuit boards (PCBs). We have used Convolutional Neural Network (CNN) and YOLOv4 for image defect detection at older vision systems that are based on predefined pixel detection. We realised three case studies of using AI with old camera setup and low computer resources, since it is often the case in real industrial systems: 1) simple image recognition, 2) recognition of the simple true or false object parameters and 3) more advanced recognition of certain defect types at the observed object. The results showed that it is possible to use AI/based solutions with older camera and computer setups, even though the time needed was significantly longer that in the case of up-to-date visual and computer resources. Application of an AI-based systems for image defect detection with older visual and computer systems reduced pseudo failures from 10% to 3%, thus making significant financial savings.

Keywords: Machine Vision · Artificial Intelligence · Quality Assurance · Small and Medium Enterprise

1 Introduction

The majority of existing small and medium enterprises (SME) have older production equipment that has no capabilities to apply Artificial Intelligence (AI) based solutions for any tasks, including quality control. Older vision systems are usually based on the predefined pixel detection [1] that commonly have lower accuracy and fewer functionalities in comparison to the new AI-based machine vision systems. A review of the literature indicates that supervised learning can provide some of the best results for computer vision applications [2].

A new direction of industry development towards smart manufacturing has engaged different AI-based solutions to assist in industrial diagnostics, whereas deep learning methods have been considered as the major support in this transition from the traditional

N. Filipović (Ed.): AAI 2023, LNNS 999, pp. 76–80, 2024.
https://doi.org/10.1007/978-3-031-60840-7_11

industry to smart diagnostics [3]. Among the deep learning methods, Convolution Neural Networks (CNNs) have been used to support image data processing, aiming at advanced industrial diagnostics, including image classification and generation, patterns prediction, data generation or security aspects in industry [3, 4]. Convolution Neural Networks (CNNs) have been widely studied for image classification and generation, including novel approaches to enhance the quality of image rendering in real-time [3].

One of the common problems in industrial settings imaging is the low light and inability to reach all the spatial and volumetric surfaces of the object that is visually scanned, thus needing new methods for data collection in the scope of designing AI-based high-quality visual systems in industry [3]. A major challenge in the high-quality processing of AI-based solutions for industrial diagnostics is acquiring comprehensive data since, in industry, available data is often limited in scope and access. Some different techniques and methods have been studied to enable data augmentation in order to increase the data amount, or to generate synthetic data that can complement data use, such as (i) basic data augmentation (DA) methods (slicing, jittering, homogenous scaling, rotation, magnitude, time and window warping, resampling, channel permutation), (ii) data augmentation through Variational Autoencoder (VAE), (iii) data augmentation through generative adversarial network (GAN), and (iv) data augmentation based on Dynamic Time Warping (DTW) [5]. Deep Neural Network (DNN) has been studied also for sentiment analysis, including the text mining and creation of libraries, such as Tensorflow or Keras, with different tools for the detection of defects or image recognition [6].

This paper presents three solutions for image processing in industrial quality control: (i) simple image classification, (ii) more complex image processing with defect descriptions and (iii) detection of the type and location of the defect. Old visual and computer systems were used to analyze the possibility of AI-based solution applications with such systems. We used Keras and TensorFlow libraries for tasks of image recognition and defect detection.

2   Materials and Methods

The case study that we analyzed in this paper focused on defect detection on printed circuit boards (PCB) which was done with an older camera and pixel detection solution, as an attempt to show how to enhance older systems and upgrade them with AI-based quality monitoring solutions. We identified 7 types of possible defects in the database: missing or misaligned components, solder bridges, solder voids, cracked PCB, overlapping traces, and contamination.

Images analyzed and processed were obtained from the 20-year-old industrial camera with a built-in light source. We selected 1000 images of good parts, and 1000 images of parts with defects. Color images of.bmp format were used. Images were preprocessed, cropped, converted to grayscale, and exported to.png format. Image contrast and brightness were adjusted, and thresholding was applied by using the Python script. The dataset was artificially expanded through the application of various data augmentation techniques such as resizing, flipping, brightness, and rotation, based on the work presented in [5]. CNN created for this application utilized Keras and TensorFlow libraries and the available tools for image recognition and defect detection [6].

Three different solutions were created. First was the simple image classification, declaring if the object is good or bad based on Keras and TensorFlow tools. The second solution declared if parts were bad and stated which type of defect was present, also based on Keras and TensorFlow tools. For these two solutions, the code had 3 convolutional layers and 2 max pooling layers. The first solution had 2 classes defined (good and bad), and the second solution had 7 classes based on 7 different defect types. The third solution detected the type and location of the defect based on YOLOv4 [7]. The third solution that used YOLOv4 architecture has a total of 52 convolutional layers. The pooling layers used in the network are max-pooling layers with a 2x2 kernel and a stride of 2. The first solution required the least amount of effort and time, and the other two required more time, and also required image labeling.

3   Results and Discussion

We analyzed the possibilities of using AI-based quality monitoring at older vision systems related to the quality check of the printed circuit boards (PCB), as shown in Fig. 1. We achieved an accuracy in the defect detection system of 90%. This led to pseudo failures, meaning that the system discarded 3% of bad parts and 10% of good parts. The older system also required adjustments in the application from time to time, like the offset of the line, or in the case of an exchange of dies. We realized data collection and recognition models, data processing and data augmentation, in order to be able to create CNN code, solution integration and its deployment. For older vision system setup (camera and computer shown in Fig. 1 on the left side) these tasks are time-consuming, also including business model aspects for determination of the most significant parameters.

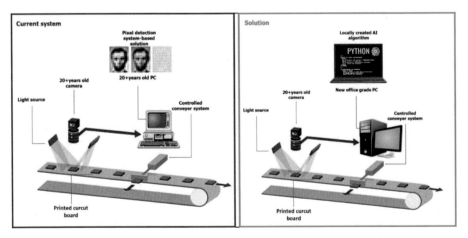

Fig. 1. Older vision system setup on the left side and AI-based quality monitoring upgrade on the right side.

One of the challenges with old machine vision systems is appearance of the pseudo failures, as well as substantially long processing times. However, the use of YOLOv4 solution to enhance the old vision system accuracy resulted in reduced pseudo failures from 10% to 3%, thus making significant financial savings. Hence model accuracy was increased from 90% to 97%. The third solution precision value was 0.92, recall of 0.95 and F1 score of 0.93. Additionally, vision system setup and adjustment time was reduced by 30%. Application of such a model in quality check showed the return on investment (ROI) after 1.5 months. If a new computer had to be bought ROI would be 2 months. And if a simple new camera (not a smart one) had to be bought alongside the computer ROI would be 2–4 months.

4 Conclusions

We analyzed possibilities to apply AI-based methods of quality control to the older vision systems that are currently mainly present in industrial settings, and especially in Small and Medium Enterprises (SMEs) with lower available resources for upgrading the process. We analyzed three case studies, from the simple image recognition, recognition of the simple true or false object parameters and more advanced recognition of certain defect types at the observed object. We used image processing by using AI-based solutions through the utilization of CNNs and YOLOv4 solutions with images acquired from the older camera types and with older computers for image processing.

Our results showed that it is possible to upgrade older vision systems with AI-powered solutions. Depending on the upgrade of the system, the return on investment can be very fast and quality assurance significantly improved. However, investment in new visual systems is recommended, due to the time-consuming tasks with older systems. Three possible solutions, from simple image recognition and to the classification of defects, all represent an improvement in the quality assurance process and promise significant savings.

Acknowledgment. This paper is funded through the EIT's HEI Initiative SMART-2M project, supported by EIT RawMaterials, funded by the European Union.

References

1. İlsever, M., Ünsalan, C.: Pixel-based change detection methods. In: Two-Dimensional Change Detection Methods. In SpringerBriefs in Computer Science. Springer, London, pp. 7–21 (2012). https://doi.org/10.1007/978-1-4471-4255-3_2
2. Hafiz, A.M., Hassaballah, M., Binbusayyis, A.: Formula-driven supervised learning in computer vision: a literature survey. Appl. Sci. **13**(2), 723 (2023). https://doi.org/10.3390/app13020723
3. Terziyan, V., Vitko, O.: Causality-aware convolutional neural networks for advanced image classification and generation. Procedia Comput. Sci. **217**, 495–506 (2023). https://doi.org/10.1016/j.procs.2022.12.245
4. Hai, J., et al.: R2RNet: low-light image enhancement via real-low to real-normal network. J. Vis. Commun. Image Represent. **90**, 103712 (2023). https://doi.org/10.1016/j.jvcir.2022.103712

5. Iglesias, G., Talavera, E., González-Prieto, Á., Mozo, A., Gómez-Canaval, S.: Data augmentation techniques in time series domain: a survey and taxonomy. Neural Comput. Appl. (2023). https://doi.org/10.1007/s00521-023-08459-3
6. Gadri, S., Chabira, S., Ould Mehieddine, S., Herizi, K.: Sentiment analysis: developing an efficient model based on machine learning and deep learning approaches. In: Vasant, P., Zelinka, I., Weber, GW. (eds.) Intelligent Computing & Optimization. ICO 2021, LNNS, vol. 371, pp. 237–247 Springer, Cham (2022). https://doi.org/10.1007/978-3-030-93247-3_24
7. Bochkovskiy, A., Wang, C-Y., Liao, H-Y.M.: YOLOv4: optimal speed and accuracy of object detection (2020). https://doi.org/10.48550/ARXIV.2004.10934

Evaluation of Nano-Object Magnetization Using Artificial Intelligence

V. A. Goranov[1]([✉]), S. Mikhaltsou[1], A. Surpi[2,6], J. Cardellini[3,4], Y. Piñeiro[5], J. Rivas[5], F. Valle[2], and V. A. Dediu[2]

[1] BioDevice Systems s.r.o., Bulharská, 10-Vršovice, 996/20, 10100 Praha, Czech Republic
info@biodevicesystems.com

[2] Istituto per lo Studio dei Materiali Nanostrutturati, (CNR-ISMN), Via Gobetti 101, 40129 Bologna, Italy
valentin.dediu@cnr.it

[3] Dipartimento di Chimica "Ugo Schiff", Università of Firenze, Sesto Fiorentino, Via della Lastruccia 3, 500129 Sesto Fiorentino, Italy
jacopo.cardellini@unifi.it

[4] CSGI, Consorzio Sistemi a Grande Interfase, Sesto Fiorentino, Via della Lastruccia 3, 500129 Sesto Fiorentino, Italy

[5] NANOMAG Laboratory (iMATUS), Universidade de Santiago de Compostela, 15782 Santiago de Compostela, Spain
jose.rivas@usc.es

[6] Istituto per la Microelettronica e i Microsistemi, IMM-CNR, 40129 Bologna, Italy
surpi@bo.imm.cnr.it

Abstract. Despite the immense potential of magnetic nanoparticles in biomedicine, their widespread application is hindered by the lack of experimental methodologies for the widely available measure of fundamental physical parameters such as hydrodynamic radius and magnetization. In this study, we propose employing "artificial intelligence-based" image analysis to extract these parameters from videos of experiments with magnetic nanoparticle concentration under the influence of magnetic field. Various solutions of magnetic nanoparticles and their complexes with liposomes, each having different magnetization levels, display distinctive temporal dynamics. The videos captured during these experiments were used to create a dataset of time-dynamic video frames, which was then processed using machine-learning techniques. For semantic segmentation of analyzed frames within a specific experiment, neural network architectures based on U-Net have been trained. We achieve quite accurate predictions of nano-object relative magnetization (RM) within a percentage error range of 11.4% to 24.1% across a spectrum of nano-object magnetizations ranging from 12 to 68 emu/g.

Keywords: machine learning · neural network · magnetic nanoparticles · magnetic field · artificial intelligence · biomedical application

© The Author(s), under exclusive license to Springer Nature Switzerland AG 2024
N. Filipović (Ed.): AAI 2023, LNNS 999, pp. 81–89, 2024.
https://doi.org/10.1007/978-3-031-60840-7_12

1   Introduction

Magnetic nanoparticles (MNPs) are considered to be an ideal platform for many biomedical applications (Gawne P.J., et al., 2023) such as targeted drug or gene delivery (Price P. M., et al., 2018) or hyperthermia (Das P., et al., 2019) because they can be manipulated remotely by external magnetic fields. The loading of biologically active agents onto MNPs offers several advantages. Indeed, this allows the delivering of therapeutic agents to selected sites, while reducing the systemic distribution of cytotoxic compounds and, ultimately, it enables effective medical treatments with lower dosage (Modena M. M., et al., 2019). Also, the loading of these agents onto MNPs limits the side effects by avoiding systemic distribution in healthy tissues (Freitas L.F. et al., 2021). Such magnetic nano-objects (MNOs) serve to protect active ingredients from rapid degradation and elimination from the body (Freitas L.F. et al., 2021). By using *in vitro* instead, MNP-loaded cells can be used for efficient manipulation in 3D magnetic scaffold (Goranov V.A. et al., 2020) for the development of the biological "tissue-like" structures.

Based on preliminary experiments involving various MNOs, the crucial parameters shaping kinetic profiles of nanoparticle motility in microfluidic systems under the influence of a magnetic field are identified as magnetization and hydrodynamic radius (Subramanian, et al., 2019). These parameters play a pivotal role in assessing the motility of MNOs in a liquid environment.

Magnetization governs how nanoparticles respond to external magnetic fields, significantly affecting their behavior in magnetic environments. This parameter is essential for designing and optimizing applications for precise control, movement, and manipulation of MNOs in diverse conditions. Understanding and fine-tuning magnetization are vital steps in enhancing the efficacy and versatility of such applications, particularly considering the heterogeneity of recently manufactured MNOs (K.M. Kosuda et al., 2016).

From the clinical point of view, for the development of special recommendations it is necessary to reduce the amount of magnetized material in MNOs-associated theranostic agents and hence the potential cytotoxicity of the magnetic core, whereas biophysical parameters of MNOs can be tailored in order to provide appropriate motility. In this context, the measurement of MNOs magnetization holds significant importance for the successful implementation of therapies relying on magnetic targeting, leading to reduced side effects and improved overall treatment outcomes in biomedical applications involving such particles. However, the accurate measurement of these characteristics through standard techniques such as magnetometry and spectroscopy is often insufficiently informative or challenging to implement outside fundamental physics laboratories. (Van De Loosdrecht M. M. et al., 2019). To tackle this problem, we used an experimental device that can be applied to the observation of readily detectable MNO's dynamic, which depends on the magnetization and external MF (at a given hydrodynamic diameter of MNO) (Cardellini J. et al., 2023).

Navier-Stokes nonlinear differential equations and convection-diffusion equations, coupled with the magnetic field equation are frequently employed to predict the dynamic behavior of the magnetic particles in fluid under external MF influence. The primary limitation of this approach lies in the challenge of considering the phase transition experienced by an MNO solution when subjected to an external magnetic field. Specifically, this transition entails a shift from a fluid of non-interacting Brownian particles to a fer-rofluid (Quanliang Cao et al. 2012). Typically, these processes are analyzed assuming unrestricted increase in the concentration of magnetic particles in regions that are charac-terized by non-uniform magnetic fields, eventually leading to their dense packing. This results in abnormally high viscosity values, typical for magnetic fluids, and the emer-gence of viscous stresses during their flow. During this phase, the magnetic particles strongly interact with both the external magnetic field and each other.

An analysis of published results (Sahai N. et al., 2021) underscores the signifi-cance of accurately measuring of MNOs accumulation parameters in confined spaces under the influence of magnetic fields, as any errors can increase significantly during mathematical analysis. Based on our experience, even when employing finite element methods, it is challenging to reliably distinguish nanoparticles with magnetization differ-ing by less than 20% through mathematical simulations of their accumulation processes (Makhaniok, A., et al., 2019).

This challenge prompts us to explore the application of artificial intelligence (AI) in the analysis of video frames from experiments involving the magnetic separation of MNOs from uniform solutions in a micro-capillary. Our findings demonstrate that this AI approach shows potential as a complementary method for assessing and predicting specific physical parameters in MNO suspensions through image analysis.

2 Methodology

The response of a uniform solution of magnetic nano-objects (MNPs or MNOs prepared using liposomes) under the influence of an external MF have been investigated by an experimental set-up described in detail in the article of Cardellini J. et al. (2023). Exper-iments on magnetization measuring have been conducted using routine methodological approaches described by Surpi A. et al. (2023). In order to simplify calculations, mag-netization values were expressed in arbitrary units. Thus, "relative magnetization" (RM) was defined as a function of MNPs amount (in content of one MNO produced using liposomes) or emu/g per MNP for standard MNPs.

Briefly, a solution of MNP (Chemicell Ltd.) or MNP-liposome hybrids (customly manufactured by Cardellini J.) are introduced into a 500 μm-wide quartz capillary and this capillary is placed into a high-gradient magnetic field generated by a couple of cubic permanent magnets. A microscope-connected CCD camera closely tracks the macroscopic response of MNOs in solution under the influence of MF. The capillary is displaced ca.200 μm from the central axis of the system to allow the concentration of the nano-objects on one side of the capillary. A dense plug of MNPs forms on the cap-illary's wall closest to the magnet and grows over time. Videos from these experiments were utilized to generate the necessary time dynamics video-frame dataset for subse-quent post-processing using machine learning. A microscope-connected CCD camera captures the process, and the resulting video frames are used to input data into the neu-ral network model. We conducted semantic segmentation of the analyzed frames in a specific experiment using neural network architectures based on U-Net (neural network set). For the neural network training and validation, we used video-frame datasets, which were extracted from the video using Python OpenCV framework and reduced to the 25 FPS and 100 × 100 pixels.

Overall, 2360 frames in 10 consecutive sets were extracted from the videos, manually marked and used as training sets. The studied areas were labeled as polygons using the LabelMe program. The region S exhibits the properties of a "magnetic fluid," where magnetic forces can accumulate nanoparticles in compact structures. In contrast, the region H is where diffusion can still affect the magnetized nano-objects and withstand the magnetic forces. The H arises from a balance between these two influences and is not observed in all cases, but crucially improves the reliability of ML results for MNOs with magnetization less than 45emu/g.

After loading from LabelMe, the annotated JSON files were converted to the standard COCO data format, resulting in files containing annotations of the areas. Semantic segmentation (Paszke A. et al., 2016; Fan Z. et al., 2023) was chosen to analyze the frames depicting the dynamics of the formation of S and H in a particular experiment. To find the optimal model for the semantic segmentation problem, we analyzed the following pre-trained neural networks (Ronneberger O. et al., 2015).

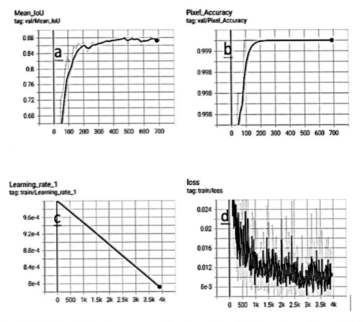

Fig. 1. Mean IOU (a) and pixel accuracy (b) of model validation. Learning rate (c) and loss function value (d) of model training using PSPNet.

UNet, ENet, PSPNet, and UNet + Resnet. The networks were fine-tuned for 700 epochs with a learning rate of 0.001 and CosineAnnealinglR for improvement of training quality (Fig. 1a). To calculate the difference between ground truth and model output we use Binary Cross Entropy loss function (H. Zhao et al., 2017); (Fig. 1b). After thorough analysis, it was found that PSPNet achieves the best validation metrics (Fig. 1c–d) – best mean intersection of unit (IOU) and pixel accuracy.

To characterize the MNOs accumulating into the visualized area of the capillary, trained PSPNet model was employed to analyze 2700 consequent images (in 7 experimental sets). The aim was to solve the inverse problem of finding the corresponding RM of particles from the data obtained through the segmentation model (Fig. 2).

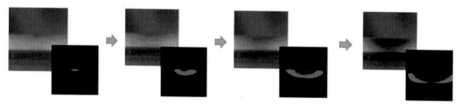

Fig. 2. Upper video-frames are the initial images and below is the neural network output.

3   Results

We analyzed the accumulation of nanoparticles with varying magnetization by examining changes in the areas of the spot and halo on the video frames. The results indicated that even slight differences in magnetization could significantly impact the dynamics of changes in abovementioned geometric parameters (Fig. 3a–b).

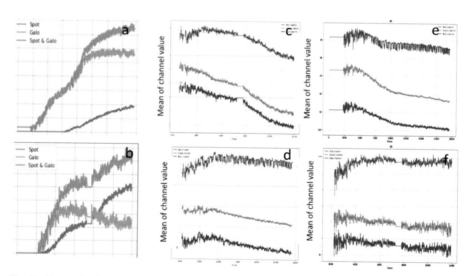

Fig. 3. Dynamic changes in the areas of the spot and halo over time for MNOs with magnetization RM48 (a) and RM54 (b). Changes in the color spectra (RGB channels) for RM48 (c) and RM54 (d) for a spot, and for the halo: RM48 (e) and RM54 (f).

Interestingly, the spectral characteristics of the nanoparticles also changed depends on RM. Notably, the color intensity along the channels also varied distinguishably corresponding to RM, even with just a 6-unit difference. The results obtained from the analysis of video frames by a neural network (displaying dynamic changes in the absolute values of the S and H areas) were processed using linear regression analysis to determine the RM (Table 1) of MNOs with a hydrodynamic radius of up to 200 nm, focusing on slow accumulation kinetics in the microfluidic capillary.

Table 1. The results of relative magnetization evaluation using described approach.

RM (MNPs/ MNO)	Evaluated magnetizations using AI approach	Approximation error (%)	RM (emu/g/MNP)	Evaluated magnetizations using AI approach	Approximation error (%)
133	104	21,8	68	73	11,4
100	83	17,0	43	78	–
54	61	12,9	24	29	24,1
48	52	11,1			

4 Discussion

The results of RM evaluation utilizing AI exhibit good consistency with the magnetization values determined during the manufacturing process. In contrast, a plethora of mathematical models commonly employed to predict the dynamic behavior of magnetic particles in fluid under the influence of an external magnetic field could lead to substantial deviations from parameters validated directly through instrumental methods (Sahai N. et al., 2021).

The primary limitation of conventional approaches lies in the inherent difficulty of accounting for the phase transition that MNPs/MNOs undergo under influence of an external magnetic field. Specifically, this transition, marked by a shift from a fluid of non-interacting Brownian particles to a ferrofluid, can result in significant alterations in liquid viscosity and MNOs aggregation (Palovics P. et al., 2020). This challenge prompted us to explore the application of AI in evaluation of video frames extracted from the video registration of MNOs behavior in uniform solutions in the microcapillary.

Among various neural network architectures, the PSPNet-based model emerged as the most suitable choice, outperforming other U-Net configurations during validation. This neural network was applied to predict the relative magnetization of MNPs and MNOs. The results clearly demonstrate the potential of AI as a complementary methodology for evaluating and predicting physical parameters within intricate systems through image analysis. Notably, it can even distinguish between MNOs that differ by a six MNPs only.

However, the applied methodology operated with MNPs may lack sufficient accuracy for objects larger than 100 nm (Hinderliter P.M. et al., 2020). This is evident in our results from calculation errors observed for nanoparticles containing 100 or more MNPs per MNP, coupled with a hydrodynamic radius greater than 100 nm.

This discrepancy may be due to the difficulty in achieving a balance between magnetic and diffusive effects, which cause difficulties that may not be effectively addressed by the methodology used. Traditionally, diffusion is inversely proportional to particle radius. Consequently, as particle sizes decrease, diffusivity increases due to heightened Brownian motion and vice versa (W. Nimisha et al., 2010). We hypothesize that this problem arises in our experiments for large MNOs due to the inability to create conditions in which magnetic and diffusive effects are in equilibrium and can accurately be captured using the applied methodology.

Ultimately, magnetic targeting can provide precise control over the transport of MNPs in microfluidic chambers. Moreover, assessing associated parameters offers the opportunity to fine-tune the delivery of therapeutic agents to specific sites, reduce systemic distribution of cytotoxic compounds, and ultimately enable more effective lower-dose medical treatments *in vivo*. In such applications, the evaluation of magnetization parameters and hydrodynamic properties of nano-(micro-) objects is crucial, but standard techniques often fall short in providing sufficient information. To address this challenge, we developed a new mathematical approach based on analyzing the macroscopic behavior of a uniform solution of MNOs under the influence of an external magnetic field followed by the AI-analysis. This approach predominantly revolves around methodology that assesses the equilibrium of magnetic field forces and diffusion effects. To the best of our knowledge, for the first time a machine learning/artificial intelligence approach was employed to predict parameters associated with the magnetization of MNOs. These mathematical models can be used to optimize and advance the development of microfluidic devices for the fabrication and characterization of engineered magnetic nanoparticles. Additionally, they can find application in other studies focused on the biological examination of nano-sized magnetized entities.

5 Conclusion

The artificial neural network PSPNet was selected to analyze the magnetic accumulation of MNOs using semantic segmentation of video frames. Leveraging the trained neural network, we achieved a high intersection over union (IOU) and pixel accuracy. In summary, employing neural networks and semantic segmentation, coupled with subsequent linear regression analysis, provides a reasonably accurate assessment of MNO magnetization. The percentage error did not exceed 24.1% for nano-objects with magnetization ranging from 17 to 68 emu/g. This study highlights the potential of neural network applications in assessing various parameters of magnetized nano-objects.

Acknowledgment. Authors express the gratitude to Debora Berti (UNIFI) for providing of MNOs for the experiments with following mathematical model development. Finally authors acknowledge the support of the EU project BOW (ID:952183).

Author Contributions. AI conceptualization - V.G.; AI software and simulation, S.M., V.G.; physical experiments, A.S.; magnetic nano-object manufacturing, J.C., J.R, Y.P.; validation, V.A.D., V.G., A.S.; formal analysis, management S.M., V.G., V.A.D., A.S., F.V.; writing—review and editing, V.G., A.S., S.M.

References

Cardellini, J., Surpi, A., Muzzi, B., Pacciani, V., Innocenti, C., Sangregorio, C., et al.: Spontaneous Formation of Magnetic-Plasmonic Liposomes with Tunable Optical and Magnetic Properties. ChemRxiv. Cambridge Open Engage, Cambridge (2023). https://doi.org/10.26434/chemrxiv-2023-q6tf3

Das, P., Colombo, M., Prosperi, D.: Recent advances in magnetic fluid hyperthermia for cancer therapy. Colloids Surf. B Biointerfaces **174**, 42–55 (2019). https://doi.org/10.1016/j.colsurfb.2018.10.051

Fan, Z., Liu, Y., Xia, M., Hou, J., Yan, F., Zang, Q.: ResAt-UNet: a U-shaped network using ResNet and attention module for image segmentation of urban buildings. IEEE J. Select. Topics Appl. Earth Observ. Remote Sens. **16**, 2094–2111 (2023). https://doi.org/10.3390/app13031493

Freitas, L.F., et al.: The state of the art of theranostic nanomaterials for lung, breast, and prostate cancers. Nanomaterials **11**(10), 2579 (2021). https://doi.org/10.3390/nano11102579

Gawne, P.J., Ferreira, M., Papaluca, M., Grimm, J., Decuzzi, P.: New opportunities and old challenges in the clinical translation of nanotheraostics. Nat. Rev. Mat. (2023). https://doi.org/10.1038/s41578-023-00581-x]

Goranov, V., Shelyakova, T., De Santis, R., et al.: 3D patterning of cells in magnetic scaffolds for tissue engineering. Sci. Rep. **10**, 2289 (2020). https://doi.org/10.1038/s41598-020-58738-5

Zhao, H., Shi, J., Qi, X., Wang, X., Jia, J.: Pyramid scene parsing network. In: 2017 IEEE Conference on Computer Vision and Pattern Recognition (CVPR), Honolulu, pp. 6230–6239 (2017). https://doi.org/10.1109/CVPR.2017.660

Hinderliter, P.M., Minard, K.R., Orr, G., et al.: A computational model of particle sedimentation, diffusion and target cell dosimetry for in vitro toxicity studies. Part Fibre Toxicol. **7**, 36 (2010). https://doi.org/10.1186/1743-8977-7-36

Makhaniok, A., Goranov, V.A., Dediu, V.A.: Determination of the protein layer thickness on the surface of polydisperse nanoparticles from the distribution of their concentration along a measuring channel. J. Eng. Phys. Thermophys. **92**, 19–28 (2019). https://doi.org/10.1007/s10891-019-01903-z

Modena, M.M., Ruhle, B., Burg, T.P., Wuttke, S.: Nanoparticle characterization: what to measure? Adv. Mater. **31**, 1901556 (2019). https://doi.org/10.1002/adma.201901556

Pálovics, P., Németh, M., Rencz, M.: Investigation and modeling of the magnetic nanoparticle aggregation with a two-phase CFD model. Energies **13**(18), 4871(2020). https://doi.org/10.3390/en13184871

Paszke, A., Chaurasia, A., Kim, S., Culurciello, E.: Enet: a deep neural network architecture for real-time semantic segmentation. arXiv preprint arXiv:1606.02147 (2016)

Price, P.M., Mahmoud, W.E., Al-Ghamdi, A.A., Bronstein, L.M.: Magnetic drug delivery: where the field is going. Front. Chem. **6**, 619 (2018). https://doi.org/10.3389/fchem.2018.00619

Cao, Q., Han, X., Li, L.: Numerical analysis of magnetic nanoparticle transport in microfluidic systems under the influence of permanent magnets. J. Phys. D: Appl. Phys. **45**, 465001 (2012)

Ronneberger, O., Fischer, P., Brox, T.: U-Net: convolutional networks for biomedical image segmentation. In: Navab, N., Hornegger, J., Wells, W., Frangi, A. (eds.) Medical Image Computing and Computer-Assisted Intervention–MICCAI 2015. LNCS, vol. 9351. Springer, Cham (2015). https://doi.org/10.1007/978-3-319-24574-4_28

Sahai, N., Gogoi, M., Ahmad, N.: Mathematical modeling and simulations for developing nanoparticle-based cancer drug delivery systems: a review. Curr. Pathobiol. Rep. **9**, 1–8 (2021). https://doi.org/10.1007/s40139-020-00219-5

Subramanian, M., Miaskowski, A., Jenkins, S.I., et al.: Remote manipulation of magnetic nanoparticles using magnetic field gradient to promote cancer cell death. Appl. Phys. A **125**, 226 (2019). https://doi.org/10.1007/s00339-019-2510-3

Surpi, A., et al.: Versatile magnetic configuration for the control and manipulation of superparamagnetic nanoparticles. Sci. Rep. **13**, 5301 (2023). https://doi.org/10.1038/s41598-023-32299-9

Van De Loosdrecht, M.M., et al.: A novel characterization technique for superparamagnetic iron oxide nanoparticles: the superparamagnetic quantifier, compared with magnetic particle spectroscopy. Rev. Sci. Instrum. **90**(2) (2019). https://doi.org/10.1063/1.5039150

Nimisha, W.: A systematic correlation of nanoparticle size with diffusivity through biological fluids. Thesis, Imperial College London (2010). http://hd.handle.net/10044/1/6080

Predicting Discus Hernia from MRI Images Using Deep Transfer Learning

Tijana Geroski[1,2]([✉]) [ID], Vesna Ranković[1] [ID], Vladimir Milovanović[1] [ID],
Vojin Kovačević[3,4] [ID], Lukas Rasulić[5,6] [ID], and Nenad Filipović[1,2] [ID]

[1] Faculty of Engineering, University of Kragujevac, Kragujevac, Serbia
tijanas@kg.ac.rs
[2] Bioengineering Research and Development Center (BioIRC), Kragujevac, Serbia
[3] Center for Neurosurgery, Clinical Center Kragujevac, Kragujevac, Serbia
[4] Department of Surgery, Faculty of Medical Sciences, University of Kragujevac, Kragujevac, Serbia
[5] School of Medicine, University of Belgrade, Belgrade, Serbia
[6] Department of Peripheral Nerve Surgery, Functional Neurosurgery and Pain Management Surgery, Clinic for Neurosurgery, Clinical Center of Serbia, Belgrade, Serbia

Abstract. The capacity to timely detect and classify discus hernia in individuals means faster access to adequate therapy. Standard way to diagnose the patients is through magnetic resonance images (MRI), which uses axial and sagittal view. Previous research has revealed that transfer learning is a useful approach when it comes to small datasets. In this paper we investigate the use of deep learning models to identify the level (healthy, L4/L5, L5/S1) and the side (healthy, bulging, left, right, center) of discus hernia in patients from MRI images. Dataset used consisted of combined publicly accessible and restricted local database of 1169 MRI images in sagittal view and 557 images in axial view. A board-certified radiologist manually classified images which was used as a golden standard. Several well-known convolutional neural networks were used in combination with transfer learning (i.e. VGG16, VGG19, DenseNet121, Xception). The results reveal competitive accuracy, as well as other metrics such as sensitivity, specificity, precision etc. Although the acquired performance is quite positive, additional investigations on larger datasets are necessary to get more robust conclusions.

Keywords: transfer learning · discus hernia · deep learning · classification

1 Introduction

Lumbar disc herniation is one of the most frequent intervertebral disc problems, resulting in restricted mobility and excruciating discomfort. A "herniated disc" is the process by which the gel-like core of a disc ruptures through a rip in the fibrous annulus. This gel stimulates the spinal neurons, causing mechanical and chemical irritation, resulting in spinal nerve inflammation and edema. More than 90% of surgical spine surgeries are performed due to disc herniation [1]. Herniated discs are more frequent in individuals

© The Author(s), under exclusive license to Springer Nature Switzerland AG 2024
N. Filipović (Ed.): AAI 2023, LNNS 999, pp. 90–98, 2024.
https://doi.org/10.1007/978-3-031-60840-7_13

aged 30 to 40, with middle-aged and older persons being somewhat more vulnerable if they participate in strenuous physical activity. Lumbar disc herniation, which occurs 15 times more commonly than cervical herniation, is a common cause of lower back pain and leg discomfort. Disc hernia is known to occur 8% of the time in the cervical region and just 1–2% of the time in the thoracic region [2].

In recent years, deep learning-based models (particularly convolutional neural networks (CNN)) have been demonstrated to outperform traditional AI techniques in most computer vision and medical image analysis applications. In cases of large datasets, deep learning methodology requires large amount of time and resources for training. To overcome these challenges, researchers have developed a transfer learning technique for medical image processing [3, 4]. In contrast to traditional machine learning algorithms that handle isolated tasks, transfer learning tries to transfer learned information from source tasks and utilize it to improve learning in targeted activities that are frequently associated to the source tasks. This is accomplished by fine-tuning pre-trained networks, which entails retraining some network components to fit the new domain. LeNet [5], AlexNet [6], [30], VGGNet [7], ResNet [8], GoogLeNet [9], DenseNet [10] and other networks are often utilized.

There are several techniques of diagnosing intervertebral disc disease in the lumbar spine. Traditional image processing approaches struggle to localize lumbar discs due to image quality, noise, and disc changes in size, shape, and appearance of vertebrae and discs. This is especially true when the disease is accompanied with a difficulty in determining the level and side of herniation. Furthermore, traditional image processing methods rely heavily on manually set parameters [11], and their success is dependent on a trial-and-error process to determine which features best describe different classes. On the other hand, CNNs have proven to be an efficient method in solving various image problems, have achieved high accuracy in the medical field, and require less fine-tuning and expert analysis [11].

As a result, deep learning has emerged as a solution. Deep learning algorithms for recognizing and segmenting intervertebral discs and vertebrae in 2D pictures or volumetric data were developed by Jackson et al. [12]. Several deep learning-based approaches for locating intervertebral discs (IVDs) and vertebrae in 2D images have been developed by authors such as Simonyan et al. [7], He et al. [8], and Dou et al. [13]. Cai et al. [14] used a 3D hierarchical model to detect IVDs and segregated them using deep neural network features. Chen et al. [15] and Suzani et al. [16] employed deep learning methods for segmentation, with Chen et al. focusing on Convolutional Neural Networks (CNN) and Suzani et al. focusing on feed forward neural networks. Harun et al. [17] created a convolutional neural network technique based on radiological ratings to distinguish between intervertebral discs and vertebrae.

When investigating these studies, it has been shown that no previous research has implemented transfer learning in multi-view, multi-class classification for the problem of discus hernia. Unlike traditional techniques of medical image classification, which require a two-step procedure (hand-crafted feature extraction + recognition), we propose an end-to-end deep learning framework based on transfer learning that predicts the disc herniation without the need for manual feature extraction. Several predefined CNN architecture designs have been used, with optimization of network hyperparameters.

2 Materials and Methods

We investigated the efficacy of transfer learning in the classification of disc hernia. The main reason for using transfer learning, as opposed to training the networks from scratch, is the limited dataset size.

2.1 Dataset

The dataset used in this study was a mix of Lumbar Spine MRI Dataset collected from Mendeley Data [18] and images obtained from 23 patients at the Clinical Centre of Kragujevac, Serbia. The use of imaging data was approved by the Ethics Committee of the Clinical Centre Kragujevac, decision number 01–11484 from September 8, 2016. As a result, number of images per class were as following (i) sagittal view – 360 healthy, 412 L4/L5 and 397 L5/S1 hernia, (ii) axial view – 243 healthy, 64 disc bulging, 75 central hernia, 110 left hernia and 65 right hernia (Table 1).

Table 1. Description of the dataset.

	Number of patients	Sagittal view			Total
		healthy	L4/L5	L5/S1	
Mendeley Data	220	329	363	318	1010
Local database	23	31	49	79	159
Total	**243**	**360**	**412**	**397**	**1169**

		Axial view					
		healthy	bulge	central	left	right	Total
Mendeley Data	220	205	56	67	85	49	**500**
Local database	23	38	8	8	25	16	**57**
Total	**243**	**243**	**64**	**75**	**110**	**65**	**557**

2.2 Proposed Methodology

The suggested workflow [19], which is similar to the broad idea of deep learning workflow and the steps performed were:

1) Region of interest (ROI) segmentation - in this stage, we used a U-net based convolutional neural network to extract the disc area in both axial and sagittal view MRI images.
2) Bounding box cropping - in this stage, we performed contour recognition and drew bounding boxes around the segmented region to narrow the search area for the CNN during the classification process.

3) ROI Enhancement - in this step, we utilized a number of image processing techniques to increase the contrast of the cropped image and highlight typical segments of the ROI for use in the diagnostic procedure.
4) Classification - the image is classified into appropriate classes (healthy, bulge, central, right or left herniation for axial view and healthy, L4/L5, L5/S1 level of herniation on sagittal view) based on transfer learning.

Figure 1 depicts the outlined idea of the proposed system.

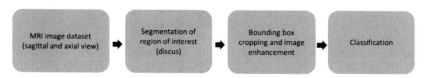

Fig. 1. Proposed workflow for disc hernia classification.

Transfer learning was used with ImageNet initial weights [20]. To diagnose disc hernia, Keras deep learning models (i.e. VGG16, VGG19, DenseNet121, Xception) were used with the following steps:

- using layers from a previously trained model with the exception of the last layer,
- freezing layers to prevent losing learned information,
- adding new and trainable layers on top of the frozen layers. These layers will be used to learn how to transform existing characteristics into predictions using the disc hernia dataset, as well as to train new layers using the disc hernia dataset.

Finally, fine-tuning was carried out, which comprised of unfreezing the whole model obtained before and re-training it on the new data with a very low learning rate (LR = 0.00001).

Standard assessment metrics such as confusion matrix, accuracy, F1 score, sensitivity (recall), specificity, and precision were employed throughout the classification stage. Furthermore, the receiver operating characteristic curve and the precision recall (PR) curve were employed.

The processing hardware were NVIDIA Quadro RTX 6000 GPU, 64GB of RAM, and a 2.40GHz Intel(R) Xeon(R) Gold, 6240R CPU. Tensorflow and Keras were used in the Python programming language to build the network.

3   Results and Discussion

Since the first step in preprocessing was segmentation, it was necessary that this is achieved with high accuracy. Table 2 shows the results for test sets for axial and sagittal view, respectively. Optimized hyperparameters for U-net model were number of epochs of 35, batch size of 8, and learning rate of LR = 0.01 [19].

Visual comparison of the original image, manually segmented mask by the doctor expert and automatically segmented disc in axial view is depicted in Fig. 2.

Table 2. Statistical measures for test set in segmentation stage

		Statistical measure				
		accuracy	dice	IOU	precision	recall
Axial view	*Test*	0.996	0.961	0.925	0.986	0.954
Sagittal view	*Test*	0.997	0.897	0.813	0.976	0.860

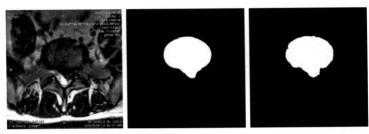

Fig. 2. Original image (left), manually segmented disc (middle) and automatic segmented disc (right) in axial view.

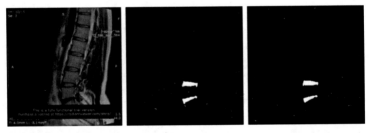

Fig. 3. Original image (left), manually segmented discs (middle) and automatic segmented discs (right) in sagittal images.

Visual comparison of original image, manually segmented mask by the doctor expert and automatically segmented disc in axial view is depicted in Fig. 3.

It can be seen that high accuracy results were achieved in segmentation stage and these results can be forwarded to the second stage of processing. The segmented region was then cropped to a size of 64x64 pixels in the second stage. This implies that all of the images have been cropped and scaled to the same dimensions. It should be noted that a 10% increase in the size of the bounding box was conducted to ensure that any segmentation mistakes did not affect subsequent classification. Furthermore, ROI enhancement employing CLAHE histogram equalization increased the visibility of the disc region from the surrounding tissue. The last step in classification was transfer learning. The results of the transfer learning have been summarized in Table 3. The metric reported was accuracy for each class. It has been shown that transfer learning delivers competitive performance when compared to the CNN built from scratch.

Table 3. Results of transfer learning (accuracy) with different base models*

Disc hernia classification (axial view)						
	healthy	bulge	central	left	right	class average
VGG16	0.91	**0.86**	0.42	0.59	0.48	0.71
VGG19	**0.95**	0.59	**0.56**	**0.64**	**0.57**	**0.75**
DenseNet121	0.93	0.57	0.46	0.53	0.40	0.68
Xception	0.93	0.54	0.20	0.44	0.15	0.62

Disc hernia classification (sagittal view)				
	healthy	L4/L5 level	L5/S1 level	class average
VGG16	**0.86**	**0.90**	**0.89**	**0.88**
VGG19	0.84	0.89	0.86	0.86
DenseNet121	0.81	0.85	0.87	0.85
Xception	0.74	0.83	0.75	0.77

*no pretrained AlexNet model is available in Keras

It should be emphasized that we have also tested the same workflow without all the steps, but only giving the network raw images as the input, but the results were not satisfying, mainly because of the large search space, that influenced the network to misclassify most of the images.

Transfer learning delivered competitive performance when compared to the competitive studies (Table 4). The main contribution of this paper is that it delivers results for both views, and it performs multiclass classification. As a consequence, the findings of this study should serve as a reference for future research focusing only on transfer learning applied to disc hernia detection.

One of the limitations of the study could be debatable if ideal hyperparameters have a broad search space (i.e., the number of last layers that will be (un)frozen, the number of last layers retrained, and the number and type of new layers added). Future research will be guided in this direction. Furthermore, pretrained weights might be loaded not only from the ImageNet dataset, but also from a dataset more similar to the dataset used in this work (i.e. model trained on MRI spine images) in order to compare the findings and investigate the possibility of negative transfer learning. Since the whole approach is automatized, the next step is validation of such methodology in clinical environment, as assistance in diagnostics.

Table 4. Overview of the best published results in similar problems. Performance metric is overall accuracy.

	Zhang et al. (2017) [21]	Jamaludin et al. (2017) [22]	Lu et al. (2018) [23]		Salehi et al. (2019) [24]	Pan et al. (2021) [25]
Investigated problem	Spinal canal stenosis	Spinal canal stenosis	Spinal canal stenosis		Disc hernia	Disc hernia
Type of scan	Axial	Sagittal	Axial + Sagittal		Axial	Axial
Number of images	582	12018	22796		2329	3555
Binary/Multi classification	binary (healthy, diseased)	binary (healthy, diseased)	Binary (healthy, diseased)	multi (normal, mild, moderate, severe)	multi (normal, bulge, protrusion, extrusion)	multi (healthy, bulge, hernia)
Performance	86.6 ± 3.3	91.4	96.3 ± 0.46	80.4 ± 1.6	87 ± 7	88.76 ± 3.72

4   Conclusions

This research proposes an automated method for identifying lumbar disc herniation using MRI axial and sagittal images. The objective was to create a decision-making system that can help physicians with diagnostic accuracy and speed. The dataset was composed of 1169 sagittal view pictures and 557 axial view images obtained from a combination of internet available datasets and locally gathered images at the Clinical Centre Kragujevac. Our technique included numerous phases. The initial stage was to use a U-net convolutional neural network to automatically find and segment L4/L5 and L5/S1 spinal discs. This was accomplished with great precision on both axial (dice = 0.961, IOU = 0.925) and sagittal (dice = 0.897, IOU = 0.813) images. The CLAHE filter was employed to improve the quality and contrast of the segmented region, and the bounding box was retrieved for use in the final classification. Based on the generated convolutional neural networks (CNNs), each cropped region containing a vertebral disc was classified into appropriate classes (healthy, bulge, central, right or left herniation for axial view and healthy, L4/L5, L5/S1 level of herniation for sagittal view). The classification accuracy was 0.75 on axial view pictures and 0.88 on sagittal view images. When compared to the accuracy of the state-of-the-art literature, our results indicated competitiveness, especially since we have dealt with multi-input and multiclass classification.

Acknowledgement. The research was funded by the Ministry of Science, Technological Development and Innovation of the Republic of Serbia, contract number [451-03-47/2023-01/200107 (Faculty of Engineering, University of Kragujevac)]. This research is also supported by the project that has received funding from the European Union's Horizon 2020 research and innovation programmes under grant agreement No 952603 (SGABU project). This article reflects only the author's view. The Commission is not responsible for any use that may be made of the information it contains.

References

1. An, H.S., Anderson, P.A., Haughton, V.M., Iatridis, J.C., Kang, J.D., Lotz, J.C.: Introduction: disc degeneration: summary. Spine 29(23), 2677–2678 (2004)
2. Cai, Y., Landis, M., Laidley, D.T., Kornecki, A., Lum, A., Li, S.: Multi-modal vertebrae recognition using transformed deep convolution network. Comput. Med. Imaging Graph. 51, 11–19 (2016)
3. Chen, H., Dou, Q., Yu, L., Qin, J., Heng, P.A.: VoxResNet: deep voxelwise residual networks for brain segmentation from 3D MR images. Neuroimage 170, 446–455 (2018)
4. Deng, J., Dong, W., Socher, R., Li, L., Li, K., Fei-Fei, L.: Imagenet: a large-scale hierarchical image database. In: 2009 IEEE Conference on Computer Vision and Pattern Recognition, pp. 248–255 (2009)
5. Dou, Q., et al.: 3D deeply supervised network for automated segmentation of volumetric medical images. Med. Image Anal. 41, 10–54 (2017)
6. Harun, N.F., Yusof, K.M., Jamaludin, M.Z., Hassan, S.A.: Motivation in problem-based learning implementation. Procedia Soc. Behav. Sci. 56, 233–242 (2012)
7. He, K., Zhang, X., Ren, S., Sun, J.: Deep residual learning for image recognition. In: Proceedings of the IEEE Conference on Computer Vision and Pattern Recognition, pp. 770–778 (2016)
8. Huang, G., Liu, Z., Van Der Maaten, L., Weinberger, K.: Densely connected convolutional networks. In: Proceedings of the IEEE Conference on Computer Vision and Pattern Recognition, pp. 4700–4708 (2017)
9. Jackson, R.P., Cain, Jr, J.E., Jacobs, R.R., Cooper, B.R., McManus, G.E.: The neuroradiographic diagnosis of lumbar herniated nucleus pulposus: II. A comparison of computed tomography (CT), myelography, CT-myelography, and magnetic resonance. Spine 14(2), 1362–1367 (1989)
10. Jamaludin, A., Kadir, T., Zisserman, A.: SpineNet: automated classification and evidence visualization in spinal MRIs. Med. Image Anal. 41, 63–73 (2017)
11. Jordan, J., Konstantinou, K., O'Dowd, J.: Herniated lumbar disc. BMJ Clin. Evid. Archiv. 2009, 1118 (2011)
12. Krizhevsky, A., Sutskever, I.E.H.G.: Imagenet classification with deep convolutional neural networks. Adv. Neural Inf. Process. Syst. 25 (2012)
13. LeCun, Y., Bottou, L.Y.B., Haffner, P.: Gradient-based learning applied to document recognition. Proc. IEEE 86(11), 2278–2324 (1998)
14. Lu, J.T., Pedemonte, S., Bizzo, B., Doyle, S., Andriole, K.P., Michalski, M.H.: Deep spine: automated lumbar vertebral segmentation, disc-level designation, and spinal stenosis grading using deep learning. In: Machine Learning for Healthcare Conference, pp. 403–419 (2018)
15. Mehrotra, R., Ansari, M., Agrawal, R., Anand, R.: A transfer learning approach for AI-based classification of brain tumors. Mach. Learn. Appl. 2, 100003 (2020)
16. Pan, Q., et al.: Automatically diagnosing disk bulge and disk Herniation with lumbar magnetic resonance images by using deep convolutional neural networks: method development study. JMIR Med. Inform. 9(5), e14755 (2021)
17. Ravì, D., et al.: Deep learning for health informatics. IEEE J. Biomed. Health Inform. 21(1), 4–21 (2016)

18. Salehi, E., Khanbare, S., Yousefi, H., Sharpasand, H., Sheyjani, O.S.: Deep convolutional neural networks for automated diagnosis of disc herniation on axial MRI. In: IEEE 2019 Scientific Meeting on Electrical-Electronics & Biomedical Engineering and Computer Science (EBBT) (2019)

19. Simonyan, K., Zisserman, A.: Very deep convolutional networks for large-scale image recognition. arXiv preprint arXiv:1409.1556 (2014)

20. Sudirman, S., Al Kafri, A., Natalia, F., Meidia, H., Afriliana, N., Al-Rashdan, W.E.: Lumbar Spine MRI Dataset (2019). https://doi.org/10.17632/k57fr854j2.2

21. Šušteršič, T., Ranković, V., Milovanović, V., Kovačević, V., Rasulić, L.F.: A deep learning model for automatic detection and classification of disc herniation in magnetic resonance images. IEEE J. Biomed. Health Inf. **26**(12), 6036–6046 (2022)

22. Suzani, A., Seitel, A., Liu, Y., Fels, S., Rohling, R.N., Abolmaesumi, P.: Fast automatic vertebrae detection and localization in pathological CT scans-a deep learning approach. In: International Conference on Medical Image Computing and Computer-Assisted Intervention (2015)

23. Szegedy, C., Liu, W., Jia, Y., Sermanet, P., Reed, S., Anguelov, D.E.: Going deeper with convolutions. In: Proceedings of the IEEE Conference on Computer Vision and Pattern Recognition, pp.1–9 (2015)

24. Yu, X., Wang, J., Hong, Q., Teku, R., Wang, S.H., Zhang, Y.D.: Transfer learning for medical images analyses: a survey. Neurocomputing **489**, 230–254 (2022)

25. Zhang, Q., Bhalerao, A., Hutchinson, C.: Weakly-supervised evidence pinpointing and description. In: International Conference on Information Processing in Medical Imaging, pp. 210–222. Springer, Cham (2017). https://doi.org/10.1007/978-3-319-59050-9_17

Prediction of Subsonic Flutter Speeds for Composite Missile Fins Using Machine Learning

Mirko Dinulović[(✉)]

Faculty of Mechanical Engineering, Aerospace Department MIT Module, The University of Belgrade, Kraljice Marije 16, 11120 Belgrade, Serbia
mdinulovic@mas.bg.ac.rs

Abstract. In the present paper, using machine learning techniques (ML), the prediction of flutter speeds for composite material missile fins is estimated for the subsonic flight regimes. Several types of composite materials are investigated (Kevlar, carbon, and E-glass S) for different fin geometries that are used nowadays in the aerospace industry as a building block for fly-worthy primary structural components. For flutter speeds data collection, the hybrid methodology is deployed, experimental, by performing tests in the subsonic wind tunnel, whereas the synthetic data for the ML model is generated using modified NACA flutter boundary equations. These equations were modified for orthotropic materials manufactured in the form of thin-walled structures. Based on this dataset, several algorithms were analyzed in order to create the ML flutter model, and it was found that for the problem on hand, the LightGBM regression approach renders the most accurate results (max coefficient of determination values) when compared to other investigated algorithms. Using ML Net technologies, a bespoke flutter software was developed. The results obtained were compared to the results obtained by experiments, known flutter binary models, and finally, numerical models based on structural, aerodynamic spline finite element approach. A good agreement between flutter speeds models was obtained, and it was concluded that the ML approach based on the LightGBM regression can be successfully used for problems where the subsonic flutter speeds estimate for composite missile fins is sought.

Keywords: flutter · stability loss · composites · fins · machine learning · wind tunnel

1 Introduction

Flutter arises from the interaction of elastic, inertial, and aerodynamic forces acting on elastic bodies and is considered a dynamic aeroelastic instability characterized by sustained vibrations of the structure. The ability of a structure to withstand these vibrations is usually expressed using flutter velocity as a measure of structural stability, representing the maximum velocity at which the structure is considered stable. Exceeding the flutter speed can destabilize the structure and lead to structural failure. Many current standards

N. Filipović (Ed.): AAI 2023, LNNS 999, pp. 99–107, 2024.
https://doi.org/10.1007/978-3-031-60840-7_14

require the flutter speed to be at least 20% greater than the "never exceed" (VNE) airspeed. It represents a very complex phenome that may lead to the loss of flying vehicle stability, or in some cases destruction of the structure itself. First structural failures due to flutter were observed in the early days of WWI where structures, that were statically compliant failed in dynamic (flight) regimes. Flutter as a phenomenon is the focus of many researchers these days and many approaches have been established from analytical to numerical with different solution algorithms. Furthermore, the flutter estimation problem has become even more complex and challenging with the introduction of composite materials into structures primarily in the aerospace industry. Many approaches developed in dynamic aeroelasticity for flutter speed estimate had to be revised in order to include the intrinsic orthotropic composite material characteristics and more complex lifting surface geometries that are now feasible from a manufacturing point of view using different composite technologies. Still, even while using modern numerical algorithms (like the finite element approach), many flutter problems need further investigation, and many fly-worthy primary structural components made of engineered composite materials are designed with very high safety factors which results in a possible overweight and not optimal solution. The loss of stability on a composite plate in subsonic flow is depicted in the following picture (Fig. 1.)

Fig. 1. Loss of stability on a composite missile fin (wind-tunnel testing).

From the previous figure bending and torsion modes of the structure are noticeable, which obviously leads to instability, control loss, and potentially structural failure.

In this paper the AI approach, which uses the ML value prediction for flutter speed estimate of composite lifting and control surfaces (like missile fins) is taken to investigate the possibilities of ML value prediction and assess the benefits of this approach compared to other "conventional methods" (analytic and numeric).

2 ML Flutter Model

As it was previously said, in this research, several objectives were sought: firstly investigate the potential application of ML techniques in composite fins missile flutter speed calculation, and recognize its potential advantages and disadvantages compared to currently applicable techniques (semi-experimental, experimental, and computational flutter analysis-CFA). Secondly, bearing in mind that the flutter of isotropic materials represents a complex phenomenon, can ML techniques be used for more intricate materials such as engineered composites which are used progressively in industry (aerospace primarily). Finally, recognize ML algorithms that can be efficiently used for the problem on hand.

In the present analysis, and flutter ML model development flowing procedure is used:

1. Problem identification and definition.
2. Data collection, experimental and synthetic data.
3. Data Preparation. Cleaning and preprocessing the data to remove noise and inconsistencies, and scaling the data to a common range.
4. Machine learning algorithm(s) analysis and selection. There are many different machine learning algorithms available, each with its own strengths and weaknesses.
5. Model Training. This involves feeding the model the input features and output targets and allowing the model to learn the relationship between the two.
6. Model evaluation. Evaluation of model performance on the test data. This will give you an idea of how well the model will generalize to new data.
7. Model deployment. Once you are satisfied with the performance of the model, you can deploy it to production. This involves integrating the model into a software application.

Composite materials have unique properties that can affect their flutter behavior. For example, composite materials are anisotropic, which means that their properties vary depending on the direction in which they are measured. This can lead to complex flutter behavior, as the direction of the aerodynamic loads can affect the flutter behavior of the fins. Additionally, composite materials can experience damage under cyclic loading, which can further complicate the flutter behavior of the fins. To overcome these challenges, researchers are developing new techniques for predicting and preventing the flutter of composite missile fins. In this paper machine learning is being used to develop models that can capture the complex interactions between the material properties, aerodynamic loads, and structural behavior of the fins. These models can be used to optimize the design of the fins and reduce the likelihood of flutter.

First, the Data set required for the ML flutter model is obtained by performing actual tests in a subsonic wind tunnel for the axial flow speeds in the range of 10–50 m/s. Several material combinations based on Kevlar, Carbon, and E-Glass are investigated.

Apart from materials, geometric characteristics of the fin surface were taken into account for the problem on hand. Recognition of geometric parameters that influence the flutter velocities was done based on previous experiences, literature overview, and by performing the DOE (design of experiments, based on Taguchi method) and sensitivity analysis of modal fin characteristics based on (Lanczos algorithm), since it is observed that frequencies in bending and torsion of the analyzed composite fin highly impact the flutter (stability loss) of the composite fin. The wind tunnel set up for experimental data collection is presented in Fig. 2.

Synthetic data in machine learning refers to artificially generated data that mimics the statistical characteristics and patterns of real-world data. It is created using various algorithms and techniques to simulate the underlying patterns and relationships present in the original data. Synthetic data can be used as a substitute or complement to real data for various purposes in machine learning, particularly when obtaining large amounts of

Fig. 2. Test Setup for the experimental portion of the dataset.

real data is difficult, expensive, or poses privacy concerns. Overall, synthetic data is a powerful tool in the machine learning toolkit, especially in situations where obtaining real data is challenging or impractical. It can accelerate model development, improve data privacy, and enhance the performance of machine learning algorithms.

Composite fins made of different types of composites (Kevlar, Carbon and E-Glass are presented in the following figure with their respective geometries (Fig. 3).

Fig. 3. Test Samples.

Hence, the parameters that were varied, in order to collect the data required for the ML model are composite material characteristics, fin thickness (t), fin aspect ratio (A), span (L), and the tip-to-root chord ratio (called taper ratio λ).

The synthetic data was required as well (since the wind tunnel tests, which are even considered as the most reliable, are expensive and demand lots of resources) to expand experimentally the obtained results. The approach to generating synthetic data was as follows: The known Naca flutter boundary equation was modified based on the Classical lamination theory, with the objective to express the Shear Modulus (G) for Isotropic materials with the equivalent shear modulus of composite laminate (G_{eq} - equivalent). In its modified form the flutter velocity based on NACA formulation for orthotropic materials can be expressed as:

$$\left(\frac{V_f}{a}\right)^2 = \frac{G_{eq}}{\frac{39.3 \cdot A^3}{\left(\frac{t}{c}\right)^3 \cdot (A+2)} \left(\frac{\lambda+1}{2}\right) \cdot \left(\frac{p}{p_0}\right)}, \tag{1}$$

In the previous equation, V_f is the flutter speed sought, c is the speed of sound at the flight altitude, p_0 is the atmospheric pressure at sea level and p is the pressure at the altitude of flight. To obtain the value of the equivalent shear laminate modulus fin, ABD matrix is computed and by twisting them D_{66} is extracted from the ABD inverse. The equivalent shear modulus is then computed from the relation:

$$
\begin{aligned}
A_{i,j} &= \sum_{k=1}^{n} Q_{i,j}^k (h_k - h_{k-1}) \\
B_{i,j} &= \frac{1}{2} \sum_{k=1}^{n} Q_{i,j}^k \left(h_k^2 - h_{k-1}^2\right) \\
D_{i,j} &= \frac{1}{3} \sum_{k=1}^{n} Q_{i,j}^k \left(h_k^3 - h_{k-1}^3\right) \\
G_{eq} &= \frac{12}{t_{lam}^3 \cdot D_{66}^{inv}}
\end{aligned}
\tag{2}
$$

This approach, in synthetic data generation is deployed (even though less accurate than experimental or CFA) as it is less computationally involved, can be easily programmed and can render large data sets faster than previously mentioned known methods with acceptable engineering accuracy. For completeness, the numerical CFA approach is presented in the following picture and requires the generation of structural, aerodynamic, and spline models (Fig. 4).

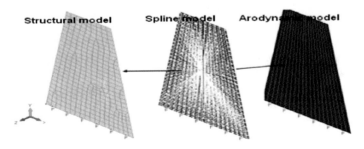

Fig. 4. CFA numerical flutter model.

The algorithms used on the previously created data set, in order to create a ML model for flutter of composite material fin, with R squared and training duration parameters are summarized in following Table 1:

The R^2 (R-squared) value is a statistical measure that represents the proportion of the variance in the dependent variable that is predictable from the independent variables

Table 1. Trainer algorithms.

	Trainer	R squared	Duration
1	FastForestRegression	0.7813	0.7500
2	SdcaRegression	0.0083	2.5980
3	FastTreeRegression	0.0706	0.3430
4	FastForestRegression	0.7813	0.3340
5	LbfgsPoissonRegressionRegression	−0.3708	0.3800
6	LightGbmRegression	0.6611	0.9340
7	LightGbmRegression	0.9988	6.3280
8	FastForestRegression	0.7962	0.3880

in a regression model. In the context of LightGBM regression (or any regression model), R^2 is used to assess how well the model fits the data.

LightGBM regression trains a model to predict a continuous numeric output given a set of input features. The model is built by iteratively adding decision trees that minimize the loss function. At each iteration, the model computes the slope of the loss function with respect to predictions and updates the model parameters to minimize the loss. LightGBM uses a special technique called "Gradient-based One-Sampling" (GOSS) to speed up the training process. GOSS samples instances with large gradients with higher probability to reduce the number of instances are considered when computing gradients. This technique helps speed up training while maintaining accuracy. LightGBM is a popular open-source gradient boosting framework that uses decision tree algorithms to perform regression and classification tasks. "Light" in LightGBM refers to the fast performance achieved through multiple techniques such as gradient-based single-sided sampling and exclusive feature bundling.

LightGBM works by building a series of decision trees, each of which is trained to predict the residual error of the previous tree. The residual error is the difference between the actual target value and the predicted target value of the previous tree (Fig. 5).

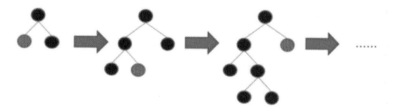

Fig. 5. LightGBM algorithm: Leaf-wise tree growth.

This algorithm uses a number of optimizations to make it faster and more efficient than other gradient-boosting algorithms. For example, LightGBM uses a histogram-based algorithm to split the data at each node of the decision tree. This is much faster than the traditional greedy algorithm, which searches through all possible split points.

It is also very scalable and it can be used to train models on very large datasets, even on distributed systems.

Finally, the complete process from problem recognition and definition to software development is presented in the following schema (Fig. 6).

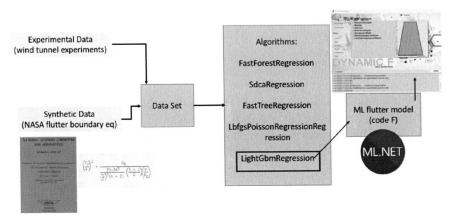

Fig. 6. ML flutter model design, data creation, algorithms analysis, and implementation.

3   Conclusions

In this work, the subsonic flutter speed model based on Machine Learning value prediction (Vf – flutter speed) is developed. By performing experimental and numerical tests it was found that the developed flutter model renders similar results (within 15% compared to experimentally obtained data), which represents acceptable engineering limits. The results obtained by performing CFA (computational Flutter Analysis, based on finite element approach) are more accurate, however, require immense resources, like experimental tests in the wind tunnel.

Based on these it can be concluded that the ML approach can be used to investigate problems in dynamic aeroelasticity (like flutter), and the results obtained using AI are acceptable from a practical engineering point of view, using very little computer resources and man-power, bearing in mind that for composite structures the safety factors (FS) are high.

References

1. Wang, Y.-R., Wang, Y.-J.: Flutter speed prediction by using deep learning. Adv. Mech. Eng. **13**(11), 1–15 (2021)

2. Brunton, S.L., Kutz, J.N., Manohar, K., Aravkin, A.Y.: Data-driven aerospace engineering: reframing the industry with machine learning. AIAAJ Aeronaut. Astronaut. **59**(8) (2021)
3. Hassanien, A.E., Darwish, A., El-Askary, H.: Machine Learning and Data Mining in Aerospace Technology. Springer (2020)
4. Meijer, M.-C.: Aeroelastic prediction for missile fins in supersonic flows. In: 29th Congress of the International Council of the Aerospace Sciences, St. Petersburg (2014)
5. Akkerman, R.: On the properties of quasi-isotropic laminates. Compos. B Eng. **33**, 133–140 (2002)
6. Martin, D.: Summary of flutter experiences as a guide to the preliminary design of lifting surfaces on missiles. NACA TN 4197 (1958)

Application of Ensemble Machine Learning for Classification Problems on Very Small Datasets

Ognjen Pavić[1]([⊠]) [iD], Lazar Dašić[1] [iD], Tijana Geroski[2,3] [iD],
Marijana Stanojević Pirković[4] [iD], Aleksandar Milovanović[1] [iD],
and Nenad Filipović[2,3] [iD]

[1] Institute for Information Technologies, University of Kragujevac, 34000 Kragujevac, Serbia
`opavic@kg.ac.rs`
[2] Faculty of Engineering, University of Kragujevac, 34000 Kragujevac, Serbia
[3] Bioengineering Research and Development Center (BioIRC), 34000 Kragujevac, Serbia
[4] Faculty of Medical Sciences, University of Kragujevac, 34000 Kragujevac, Serbia

Abstract. Machine learning is one of the most widely used branches of artificial intelligence in recent years. It is most commonly used for solving classification or regression problems through the utilization of supervised learning approaches. Machine learning models require high quality and a sufficient quantity of data to produce good results. This paper investigates an approach which incorporates ensemble learning through the aggregation of multiple machine learning models for the purposes of increasing prediction capabilities in cases in which a very limited amount of data is available for training. The ensemble model was trained on a patient fractional flow reserve biomarker dataset and with the goal of classifying patients into risk classes based on their risk of suffering an acute myocardial infarction. The ensemble model was comprised of multiple random forest classification models which were trained with different combinations of training and test data to improve the prediction accuracy over the use of a single random forest model. Final ensemble achieved a prediction accuracy of 71.3% which was an immense improvement over the 36% prediction accuracy of a single random forest classification model.

Keywords: Machine learning · Classification · Risk assessment · Random forest · Ensemble First Section

1 Introduction

Machine learning is a branch of artificial intelligence which utilizes a variety of learning algorithms for the creation of mathematical or logic-based models, with a goal of representing connections and dependencies within a dataset, so that they can be utilized in the future on new data points. The most common use of machine learning is solving classification and regression problems through supervised learning approaches. Classification and regression machine learning algorithms are often used for but are not limited

N. Filipović (Ed.): AAI 2023, LNNS 999, pp. 108–115, 2024.
https://doi.org/10.1007/978-3-031-60840-7_15

to solving problems including the need for semantic classification, risk stratification, uncovering hidden knowledge, uncovering trends over time, future prediction of values through data extrapolation etc., and along with deep learning approaches are most commonly utilized in the fields of medicine, economics, computer vision and robotics.

Machine learning is a powerful tool that can be used for automation and streamlining of complex tasks and for the uncovering of hidden knowledge which can be utilized for further research in the future. However, machine learning requires data that is descriptive enough for a model to be created while also being large enough to minimize errors efficiently and avoid model underfitting. A machine learning model trained with bad data cannot produce satisfying results; on the other hand, an abundance of data is not always available for model training. In this paper we have created an ensemble of machine learning models whose purpose was to increase the predictive capabilities in contrast to using a single model with a very small amount of available data.

Ensemble learning implies the creation of multiple less complex machine learning models which work together to come to a conclusion [1]. Ensemble machine learning models can be created by aggregating multiple models of lesser complexity created by the same algorithm, but can also consist of models created using several different algorithms [2].

Cardiovascular diseases are responsible for significant medical, social and economic consequences around the world. Cardiovascular diseases one of the leading causes of disability, loss of ability to work and premature mortality and put great costs on the healthcare system. According to literature data, cardiovascular diseases result more than 30% of all deaths in the world, and half of these outcomes are the result of ischemic heart diseases [3].

Acute myocardial infarction is manifested through the necrosis of the heart muscle that occurs due to coronary artery occlusion and insufficient oxygenation of cardiomyocytes [4]. Considering such serious consequences of this disease, it is necessary to develop strategies for the prevention and early detection of cardiovascular risks, as well as for rapid diagnosis of acute myocardial infarction for the timely application of adequate therapy [5].

Fractional flow reserve (FFR) measurement is used to assess the severity of coronary artery stenosis identified during invasive coronary angiography [6]. FFR is the ratio between the maximum possible blood flow in the diseased coronary artery and the possible maximum blood flow in the normal coronary artery. This ratio shows the potential decrease in flow distal to the coronary stenosis. In healthy people, the FFR is 1, whereas an FFR lower than 0.75-0.8 indicates myocardial ischemia. FFR values less than 0.75 indicate the need for revascularization [7].

2   Materials and Methods

2.1   Dataset

The dataset consisted of data gathered from patients in the form of clinical biomarkers and demographic data, as well as descriptive data which contained information on the primary and follow up diagnosis and definitions of location and degrees of stenosis and lesions in the right coronary artery, left anterior descending artery and the left circumflex artery. The dataset also contained the measured values of FFR which were used for the creation of the classification target. The dataset contained information about 112 patients, however only 17 of these patients had known ground truth values for FFR.

The main goal was to create a machine learning model that is capable of classifying patients into appropriate risk classes based on their risk of suffering acute myocardial infarction (AMI), utilizing the patients for whom the class was known and apply the classification on the rest of the available data. In order to classify patient risk of AMI, class labels needed to be created based on the known values of measured FFR. Patients with an FFR greater than 0.8 were labeled as low-risk, while the patients with an FFR lower than 0.75 were labeled as high-risk. Patients whose FFR is between 0.75 and 0.8 are sometimes treated as belonging to a third, borderline class, which can be treated as both low-risk and high-risk and the final decision is the medical professionals prerogative [8]. For the purposes of this study, the borderline class was treated as a part of the high-risk class.

The available data had multiple problems which needed to be addressed during data preprocessing. The set contained instances of missing values which were filled in using conventional approaches depending on the type of missing data. Namely, numeric data was filled in using the mean value of the appropriate column and binary data was filled in using the most common binary value contained within that column. The second problem which existed within the data was the descriptive nature of certain features. Descriptive data containing information on stenosis and lesion in coronary arteries needed to be translated into a numeric value. This translation was conducted as following:

- Data that contains percentile values for the narrowing of the observed artery was translated as a numeric sample corresponding to the percentage value.
- Data that contains an approximation of the narrowing in the form of a range of values was translated as a numeric sample corresponding to the average value of the observed range.
- Data that does not contain percentage values of the narrowing, but has the indication that the narrowing is not substantial was translated as if it held information about a narrowing of 10%
- Data that does not contain percentage values of the narrowing, but has the indication that the narrowing is very minor was translated as if it held information about a narrowing of 5%
- Data that does not contain percentage values of the narrowing but indicates an orderly arterial lumen was translated as if there was no narrowing at all.
- Data that does not contain any indication of the size of the narrowing, nor does it contain the previously mentioned phrases with which the narrowing was estimated was not translated at all but was approximated as a mean value of all of the other translated values.

2.2 Individual Model Training

A singe random forest classification model was trained and tested using the entire available dataset. After fine tuning of the parameters, the model achieved 36% classification accuracy. For this reason, a greater ensemble model needed to be created.

Individual machine learning models were constructed by using the random forest classification algorithm. Because 17 labeled data samples were not enough to build a comprehensive test set, multiple random forest models were trained using different combinations of training and test data.

Random forest algorithm produces classification models which are ensemble models on their own, by training multiple decision tree models using randomly selected sub-samples of training and test data for each one. Each of the created decision tree models comes to its own conclusion and the random forest model chooses the output value which is present in the highest number of cases. Each random forest model was created using 50 decision trees without constraint with regards to minimum samples required for creating branching nodes and leaves.

Similarly to the final decision making process of the random forest algorithm, our created model counts the number of times each decision is made at the output of each lower-level classification model, and chooses the most numerous class label as the final model output.

2.3 Ensemble Model Creation

The available dataset was split into every possible configuration of training and test data with a 16:1, 15:2, 14:3 and 13:4 train-test split. For every existing combination of training and test data, a single random forest classification model was trained. Although many classification models had to be trained, it was not efficient for every one of them to be a part of the final ensemble, because the decision-making process would take a very long time. Only certain models, which performed the best on their own combinations of data, were selected to be the part of the final ensemble. The selection criteria were set based on their achieved prediction accuracy and the number of data samples used for testing. Models which were tested using a single data sample were selected if they achieved 100% prediction accuracy. Models evaluated using 2, 3 or 4 test samples were selected if they achieved a minimum of 50%, 66% and 75% prediction accuracy respectively. The final ensemble model was then constructed through the sequential use of the selected models. The system creation methodology is shown in Fig. 1.

3 Results and Discussion

In the original dataset, there existed a large imbalance between patients who belong to the low-risk class and patients who belong to the high-risk class. There is also a big underlying risk of falsely classifying high-risk patients into the low-risk class. For these reasons, the original evaluation method for the ensemble model was chosen to be the F1 score metric on the high-risk class. However, F1 score could not be used as the main evaluation method for multiple reasons, the most important of which being the inability to produce meaningful results if only one class is present in the test set. Namely, during lower-level model creation for the final ensemble, every combination of available labeled samples was used for training and testing, some of which did not include any samples of the high-risk class in the test set. In these situations, results produced by the application of F1 score calculation would be meaningless and the model could not be evaluated properly.

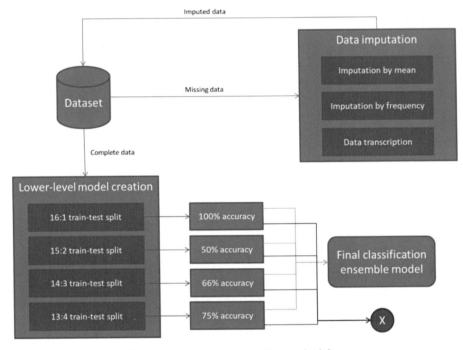

Fig. 1. Proposed system creation methodology

The final ensemble model was evaluated using the prediction accuracy metric, since prediction accuracy could be universally calculated, and did not have any constraints with regards to test set contents. The final model achieved a prediction accuracy of 71.32% which was much higher than the accuracy achieved by using a single random forest model which was at 36% after fine tuning.

Average performances for models trained with different configuration of training and test sets are shown in Table 1.

Table 1. Classification accuracy metrics

train:test split	Mean prediction accuracy
16:1	100%
15:2	56%
14:3	69%
13:4	76%
Final model	71.32%

The results achieved after utilizing an ensemble model approach, even though much better than those achieved using a single machine learning model, are not good enough to justify the use of the model in its current state as a decision support system. As it stands, there are no possible substantial improvements to prediction capabilities without the increase in the size of the dataset.

The Support Vector Machine algorithm is a kernel-based classifier, which divides the training data using multidimensional hyperplanes. The dimensionality of these hyperplanes is dependent on the dimensionality defined by the model input parameters. It is capable of perfectly separating the dataset based on training data samples while keeping the Euclidean distance between the physical representations of training data points in multidimensional space at a maximum. However, problems arise with the generalization capabilities of support vector machine models for newly introduced data. For this reason, a coefficient of error tolerance C is introduced, which allows the algorithm to make minor mistakes during training but increase the potential to better generalize when making decisions in the future [9]. The problem with our available dataset arises when any high-risk patient is present in the test set, reducing the number of available high-risk patients for training. Any value of allowed error tolerance renders the models incapable of predicting the high-risk class in an acceptable manner.

Similarly, the K-Nearest Neighbors algorithm, while not an algorithm that creates a mathematical model is still capable of separating a multidimensional hyperspace into sections belonging to attached class labels. The data sample separation of this algorithm is based on the proximity of similar training data points in the n-dimensional space. High-risk patients define two dense clusters within the aforementioned space clearly separating the zones within which the patient is considered to be under a high risk of AMI from those in which he would be considered to be under a low risk of suffering an AMI. Introducing high-risk samples into the test set reduces the density of these clusters which greatly reduces the space inside the n-dimensional hyperspace within which the patient will be classified as belonging to the high-risk class.

However, the introduction of new data in the future opens up the possibility of using multiple different classification algorithms in the construction of the final ensemble model. With the possibility of introduction of models trained with different classification algorithms the versatility and robustness of the model increases drastically along with the increase in generalization capabilities. Additionally, the introduction of new data to the training dataset increases the variety of feature values and the possible combinations of feature values which better defines the classes and improves prediction results.

In future research, 3D finite element model numerical calculations combined with real measurements of FFR could be used to significantly increase the size of the dataset and achieve better accuracy of the proposed ML models. In addition to the improvement of the proposed machine learning approach to assessing the risk of AMI, additional increase in the amount of available data would enable the transfer from machine learning algorithms to creating a specialized neural network for patient classification.

4 Conclusions

Creating machine learning models to solve classification or regression type problems will yield less than satisfying results if the dataset used for their training does not contain high quality and a high quantity of data.

In this paper we were met with the challenge of a very small dataset available for model creation. Although this challenge is impossible to overcome without using a better suited dataset, we were able to majorly improve the final results. Results were improved through the creation of a large number of less complex machine learning models and their aggregation into an ensemble model.

The final ensemble model was constructed through a selection process where lower level models needed to achieve an accuracy of 100%, 50%, 66% and 75% prediction accuracy for models tested with 1,2,3 and 4 samples respectively. In the end the final ensemble achieved 71.32% classification accuracy, which was a result far greater than the 36% classification accuracy achieved by a single random forest model trained and tested using the entire dataset.

There exist multiple other approaches to creating an ensemble model mainly through the use of multiple different machine learning algorithms during the creation of lesser complexity models, and through the change in the inner workings of the final decision selection process, but the described approach achieved a major increase in overall model performance.

In conclusion we managed to achieve satisfying results through the use of ensemble learning and improve the quality of model outputs when compared to the use of a less complex model. The final results cannot be further improved through model tuning without the expansion of the available dataset through the introduction of more labeled data.

Acknowledgement. The research was funded by the Ministry of Science, Technological Development and Innovation of the Republic of Serbia, contract number [451-03-47/2023-01/200378 (Institute for Information Technologies Kragujevac, University of Kragujevac)]. This research is also supported by the project that has received funding from the European Union's Horizon 2020 research and innovation programmes under grant agreement No 952603 (SGABU project). This article reflects only the author's view. The Commission is not responsible for any use that may be made of the information it contains.

References

1. Polkar, R.: Ensemble learning. In: Ensemble Machine Learning, vol. 1, no. 1, pp. 1–34. Springer (2012)
2. Sagi, O., Rokach, L.: Ensemble learning: a survey. In: WIREs Data Mining and Knowledge Discovery, vol. 8, no. 4 (2018)
3. GBD 2017 Causes of Death Collaborators. Global, regional, and national age-sex-specific mortality for 282 causes of death in 195 countries and territories, 1980–2017: a systematic analysis for the Global Burden of Disease Study 2017. Lancet **392**(10159), 1736–1788 (2018)
4. Thygesen, K., et al.: Fourth universal definition of myocardial infarction. Circulation **138**, 20 (2018)
5. Chan, D., Leong, L.: Biomarkers in acute myocardial infarction. BMC Med. (2010)
6. Pijls, N.H., et al.: Measurement of fractional flow reserve to assess the functional severity of coronary-artery stenoses. N. Engl. J. Med. **334**, 1703–1708 (1996)
7. Wc Lo, E., Menezes, L.J., Torii, R.: On outflow boundary conditions for CT-based computation of FFR: examination using PET images. Med. Eng. Phys. **76**, 79–87 (2020)

8. Modi, B.N., et al.: Revisiting the optimal fractional flow reserve and instantaneous wave-free ratio thresholds for predicting the physiological significance of coronary artery disease. Circulat.: Cardiovasc. Invent.. **11**(12) (2018)
9. Monien, K., Decker, R.: Strengths and weaknesses of support vector machines within marketing data analysis. Innov. Classif. Data Sci. Inf. Syst. 355–362 (2005)

An Unsupervised Image Segmentation Workflow for Extraction of Left Coronary Artery from X-Ray Coronary Angiography

Lazar Dašić[1,2]([✉]) [ID], Ognjen Pavić[1,2] [ID], Tijana Geroski[2,3] [ID],
and Nenad Filipović[2,3] [ID]

[1] Institute for Information Technologies, The University of Kragujevac, Jovana Cvijića bb,
34000 Kragujevac, Serbia
{lazar.dasic,opavic}@kg.ac.rs
[2] Bioengineering Research and Development Center (BioIRC), Prvoslava Stojanovića 6, 34000
Kragujevac, Serbia
{tijanas,fica}@kg.ac.rs
[3] Faculty of Engineering, The University of Kragujevac, Sestre Janjić 6, 34000 Kragujevac,
Serbia

Abstract. Coronary heart disease (CHD) is a serious cardiovascular illness that is among the top causes of death worldwide. Using X-ray coronary angiography, it is possible to detect and monitor CHD by visualizing coronary vessels. One of the most important steps in analyzing angiographic images is image segmentation, where the coronary arteries are separated from the background. In this paper, we propose an unsupervised image segmentation workflow that uses different filters in order to minimize the limitations of X-ray coronary angiography and achieve satisfactory segmentation of the left coronary artery. During the preprocessing step, the X-ray angiographic image of the coronary artery is processed with CLAHE, the Wiener filter and gamma correction in order to overcome the shortcomings of the X-ray imaging data. These preprocessing steps greatly reduce the background noise and improve the separation of the artery from the rest of the image. The preprocessed image is then segmented using Otsu's thresholding method, which results in a binarized image. This image has left coronary artery successfully segmented, but unfortunately a lot of non-vessel segments have been wrongly labeled as well. In the postprocessing step, connected components are obtained, and then using information about their size the largest connected component represents a segmented left coronary artery, while the rest is marked as background.

Keywords: left coronary artery · unsupervised image segmentation · x-ray coronary angiography

1 Introduction

Coronary heart disease (CHD) is a serious cardiovascular illness that is among the top causes of death worldwide, making the early detection and treatment of the disease of utmost importance [1]. Coronary arteries are small blood vessels branching from the

N. Filipović (Ed.): AAI 2023, LNNS 999, pp. 116–122, 2024.
https://doi.org/10.1007/978-3-031-60840-7_16

aorta and they supply the heart with oxygen-rich blood [2]. This disease is caused by a plaque creation that results in wall stiffness and lumen reduction, making the coronary arteries unable to supply blood to the heart. In order to diagnose and monitor CHD, various imaging modalities are widely used (echocardiogram, EKG, etc.). Most commonly used method of assessing CHD is the invasive X-ray coronary angiography and is still considered the gold standard [3]. X-ray coronary angiography provides visualization of blood vessels and their anatomical structure by utilizing X-rays during the injection of radiopaque contrast material [4]. Unfortunately, X-ray coronary angiography is not a perfect method since a lot of information can be lost due to imaging artifacts, wrongly chosen projection angles, vessel overlap, etc. [5]. Another problem is the intraobserver variability that could lead to a wrong assessment of lesion severity and stent dimensions [6]. Because of the aforementioned shortcomings, there have been numerous research efforts to use angiographic images as means for 3D reconstruction that would give better insight into the anatomical structure of the coronary arteries.

Over the years, numerous approaches for 3D reconstruction of coronary arteries that differ drastically in both technology and workflow have been developed. However, one common step for the majority of these approaches is the segmentation of coronary arteries from digital angiography images. The process of image segmentation consists of separating the region of interest from the background. Various unsupervised, as well as supervised and deep learning ideas have been developed for the angiographic image segmentation. Convolutional neural networks (CNN) achieved great results in many image segmentation tasks, so it was only natural that numerous CNN architectures have been developed for the task of coronary artery segmentation [7–9]. Even though deep learning approaches achieve great results, they require large amount of labelled data that is often not available. As a way of overcoming the need for labelled data, unsupervised learning methods have been developed. Most of the developed unsupervised approaches have been based on the use of different types of image processing filters [10–12]. These approaches achieved suboptimal results due to the aforementioned shortcomings of X-ray coronary angiography as well as the complexity of coronary vessels (especially the left coronary artery with larger number of bifurcations and small blood vessels).

2 Methods

In this work, we propose an unsupervised image segmentation workflow that uses different filters in order to minimize limitations of X-ray coronary angiography and achieve satisfactory segmentation of the left coronary artery.

2.1 Dataset

Dataset used in this research consists of X-ray coronary angiography imaging data from 147 different patients, collected during medical checkups at the Clinical Centre of Kragujevac. During the procedure, the patients' coronary arteries have been imaged from different angles in order to give better insight into the vessel structure. The collected dataset is provided in the standard DICOM format. For the methodology development, only the X-ray images that are in the end-diastole and that shows the whole coronary tree have been selected. These frames are considered as the best option for both diagnosis by medical experts and analysis by computer methods [13].

2.2 Workflow

All the steps of the developed methodology are shown in Fig. 1. The methodology consists of three main parts: preprocessing, filtering and segmentation and postprocessing.

Since the DICOM format, besides imaging pixel data, contains additional information and metadata, it was important to convert data into a format that is easier to work with, such as PNG. The extracted PNG images contained large black borders that did not contain any relevant information, so the images were cropped. An example of the left coronary artery is shown in Fig. 2a.

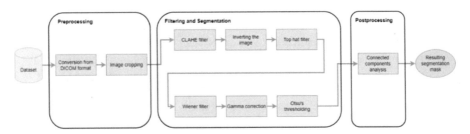

Fig. 1. Proposed methodology workflow.

One of the biggest problems working with any X-ray based imaging data is the low contrast in the image, which makes segmentation of the region of interest from the background challenging. In the case of available angiography data, coronary arteries are not distinct enough from the background muscle and bone tissue. In order to improve contrast, the grayscale image of the left coronary artery is processed using the Contrast Limited Adaptive Histogram Equalization (CLAHE) filter. CLAHE operates on small regions in the image and introduces a clipping limit of the histogram, which results in reduced over-amplification of the contrast in the image. The results of the application of the CLAHE filter are shown in Fig. 2b.

Following the CLAHE filtering, pixel values in the image are inverted so the coronary artery is represented with white color and background is represented in dark color. With the goal of suppressing the background in the images, top hat filter that enhances bright objects of interest is applied. Even though CLAHE filter significantly improved contrast in the image, it also made the image grainy and the background bone structure more prominent, as shown in Fig. 2b. In order to suppress the noise that is present in the picture a Wiener filter has been applied. Wiener filter is an adaptive low-pass filter that applies varying smoothing based on a local mean and variance of the pixels in the patch of size M × N. Wiener filters are usually applied in frequency domain, where the resulting image represents a product between a Wiener filter and an image spectrum.

In order to further enhance the separation between blood vessels and the background, the Wiener filter-processed image underwent a gamma correction procedure. Gamma

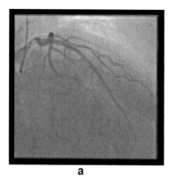 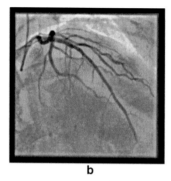

a
b

Fig. 2. Original grayscale image of left coronary artery (a); Results of application of the CLAHE filter (b).

correction is a nonlinear operation used to encode or decode luminance in the image, thus changing the saturation of the image. The level of saturation is controlled by the γ parameter, where $\gamma < 1$ is called encoding gamma and results in enhancing of saturation in the image, while $\gamma > 1$ is called decoding gamma and results in lowering of saturation in the image. For the task of coronary artery segmentation, encoding gamma of 0.97 was selected in order to slightly saturate the image and make bright blood vessels more prominent. It would have been better if the image was more saturated, but this would result in the amplification of the background noise as well.

The pixel values in the processed image are in the range of 0 to 255. For binary segmentation, we want to group all of the pixels that belong to the coronary artery and label them with one pixel value and mark the other pixels with 0. To achieve binary segmentation, a thresholding technique was used where the image is binarized based on pixel values. If the intensity of a pixel is greater than the given threshold, the pixel is marked as white and represents coronary artery region. If the input pixel value is less than the given threshold, the pixel is marked as black (background). It is crucial to decide what the optimal threshold is and it can drastically affect the quality of resulting segmented image. In order to find the optimal threshold value Otsu's thresholding algorithm was used. This algorithm finds threshold value by minimizing intra-class intensity variance, or equivalently, by maximizing inter-class variance [14].

Looking at the results of Otsu's thresholding, it can be seen that it adequately segments coronary artery region, but in the process, it also wrongly classifies some of the background areas as region of interest. These wrongly labelled areas are considered noise and have to be removed in order to correctly finalize process of the binary segmentation. Due to the small size of these areas, the connected components analysis can be utilized for the noise removal. Every connected area in the image is marked as a single component and all of the components are sorted by their size (number of pixels in each component). The largest component is marked as the coronary artery, while the rest of the segments are considered background. Figure 3 shows the whole methodology of the coronary artery segmentation.

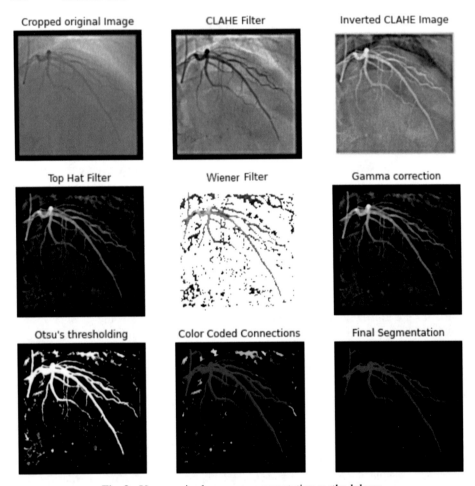

Fig. 3. Unsupervised coronary segmentation methodology.

3 Results

Figure 4 shows the final results of the image segmentation workflow on examples of left coronary artery images. It is clear that methodology is capable of achieving great segmentation results even on the smaller arteries that are hard to be correctly segmented even by medical experts.

Since the proposed methodology is considered to be an unsupervised learning technique, there is not any available labelled data that can be compared with resulting segmentation mask. This results in the lack of any quantitative metrics that could show segmentation performance of the developed methodology.

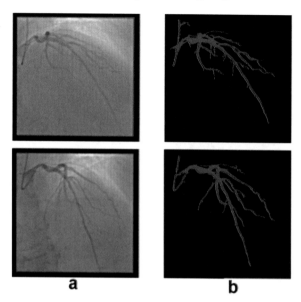

Fig. 4. X-ray angiographic images of left coronary artery (a); Result of proposed segmentation workflow (b).

4 Conclusions

In the field of medicine, segmentation of certain region of interest for diagnostic purpose is a common problem. The task of image segmentation is especially difficult when working with X-ray images that suffer from imaging artifacts and noise. In this paper we describe robust unsupervised segmentation method that extracts left coronary artery from the angiographic imaging data.

The proposed workflow correctly segments major branches of the left coronary artery as well as smaller diagonal branches. The achieved results are satisfactory, even with the previously mentioned shortcomings of X-ray images. The proposed methodology showed that it is possible to segment left coronary artery even when there is not any labelled data. Future research is going to focus on adapting the methodology for the task of segmentation of right coronary artery and the usage of the proposed workflow in a larger project that focuses on the 3D reconstruction of coronary arteries.

Acknowledgments. The research was funded by the Ministry of Science, Technological Development and Innovation of the Republic of Serbia, contract number [Agreement No. 451-03-47/2023-01/200378]. This research is also supported by the project that has received funding from the European Union's Horizon 2020 research and innovation programme under grant agreement No 952603 (SGABU project). This article reflects only the author's view. The Commission is not responsible for any use that may be made of the information it contains.

References

1. World Health Organization. World Health Statistics Overview 2019: Monitoring Health for the SDGs, Sustainable Development Goals. World Health Organization (2019)
2. Acharya, R., Hagiwara, Y., Koh, J.E.W., Oh, S.L., Tan, J.H., Adam, M., et al.: Entropies for automated detection of coronary artery disease using ECG signals: a review. Biocybernet. Biomed. Eng. **38**(2), 373–384 (2018)
3. Narula, J., Chandrashekhar, Y., Ahmadi, A., Abbara, S., Berman, D.S., Blankstein, R., et al.: SCCT 2021 expert consensus document on coronary computed tomographic angiography: a report of the society of cardiovascular computed tomography. J. Cardiovasc. Comput. Tomograph. **15**(3), 192–217 (2021)
4. Scanlon, P.J., Faxon, D.P., Audet, A.M., Carabello, B., Dehmer, G.J., Eagle, K.A., et al.: ACC/AHA guidelines for coronary angiography: a report of the American College of Cardiology/American Heart Association Task Force on Practice Guidelines (Committee on Coronary Angiography) developed in collaboration with the Society for Cardiac Angiograph. J. Am. College Cardiol. **33**(6), 1756–1824 (1999)
5. Meijering, H.W.: Image Enhancement in Digital X-ray Angiography. Ponsen & Looijen (2000)
6. Gollapudi, R.R., Valencia, R., Lee, S.S., Wong, G.B., Teirstein, P.S., Price, M.J.: Utility of three-dimensional reconstruction of coronary angiography to guide percutaneous coronary intervention. Catheteriz. Cardiovasc. Intervent. **69**(4), 479–482 (2007)
7. Tao, X., Dang, H., Zhou, X., Xu, X., Xiong, D.: A lightweight network for accurate coronary artery segmentation using x-ray angiograms. Front. Publ. Health **10**, 892418 (2022)
8. Iyer, K., Najarian, C.P., Fattah, A.A., Arthurs, C.J., Soroushmehr, S.R., Subban, V., et al.: Angionet: a convolutional neural network for vessel segmentation in X-ray angiography. Sci. Rep. **11**(1), 18066 (2021)
9. Kaba, Ş, Haci, H., Isin, A., Ilhan, A., Conkbayir, C.: The application of deep learning for the segmentation and classification of coronary arteries. Diagnostics **13**(13), 2274 (2023)
10. Frangi, A.F., Niessen, W.J., Vincken, K.L., Viergever, M.A.: Multiscale vessel enhancemenet filtering. In: Medical Image Computing and Computer-Assisted Intervention. MICCAI 1998. Springer, Heidelberg (1998). https://doi.org/10.1007/BFb0056195
11. M'hiri, F., Duong, L., Desrosiers, C., Cheriet, M.: Vessel walker: coronary arteries segmentation using random walks and hessian-based vesselness filter. In: 2013 IEEE 10th International Symposium on Biomedical Imaging (2013)
12. Yin, Z.X., Xu, H.M.: An unsupervised image segmentation algorithm for coronary angiography. BioData Mining **15**(1), 27 (2022)
13. Dehkordi, M.T.: Extraction of the best frames in coronary angiograms for diagnosis and analysis. J. Med. Signals Sens. **6**(3), 150 (2016)
14. Otsu, N.: A threshold selection method from gray-level histograms. IEEE Trans. Syst. Man Cybernet. **9**(1), 62–66 (1979)

Artificial Intelligence in Intelligent Healthcare Systems–Opportunities and Challenges

Anita Petreska$^{(\boxtimes)}$ (ORCID) and Blagoj Ristevski (ORCID)

Faculty of Information and Communication Technologies - Bitola, University "St. Kliment Ohridski", Bitola, Republic of North Macedonia
`{petreska.anita,blagoj.ristevski}@uklo.edu.mk`

Abstract. Artificial intelligence (AI) has the potential to revolutionize healthcare by improving the accuracy, speed and efficiency of biomedical systems by providing intelligent solutions that enable healthcare providers to make smart decisions and generate new insights and discoveries. Given the exponential increase in big medical data, existing classic hospital information systems are not adequate for data analysis. Artificial intelligence plays a key role in remote patient monitoring and telemedicine, especially in underdeveloped areas or during emergencies using virtual consultations. Biomedical systems use machine learning (ML) and deep learning (DL) algorithms that enable more efficient analysis of large amounts of health data, including electronic health records, medical images, and patient histories. By analyzing patient data from a variety of sources: electronic health records, real-time patient registries, and research articles, machine learning algorithms can identify changes in a patient's condition and alert healthcare providers to potential problems before they become serious. The integration of artificial intelligence (AI), machine learning (ML), data mining (DM), and data integration has ushered in new technology for transformative healthcare.

Keywords: artificial intelligence · machine learning · biomedical system · medical big data · algorithms

1 Introduction

The "Covid-19" pandemic proved that the traditional health system and doctors cannot cope with the new situation alone. The role of digital technologies and artificial intelligence is to enable the revolutionary transformation of health systems. Artificial intelligence [1] is an interdisciplinary science that, with the help of powerful computer algorithms as a result of the rapid development of machine learning, helps computers to think and learn like humans. Artificial intelligence research relies heavily on logic, knowledge, planning and communication. Artificial intelligence provides reliable evidence-based recommendations and enables patients to be actively involved in their treatment. Section 2 covers medical big data [2] in healthcare as an invaluable tool for proactive intervention and predictive analysis of patient's conditions in the process of making smart decisions. The "10 Vs" characteristics of big data, predictive, preventive, personalized and participatory (P4P) medicine, electronic health records (EHRs),

N. Filipović (Ed.): AAI 2023, LNNS 999, pp. 123–143, 2024.
https://doi.org/10.1007/978-3-031-60840-7_17

and commonly used datasets are presented. Section 3 gives an overview of analytics in the field of large amounts of medical data, the role of machine learning in healthcare, data mining, analysis of the most commonly used ML algorithms in healthcare [3, 4]. Section 4 presents machine learning in the field of large amounts of medical data [5]. In Sect. 5, a comparative analysis is made between the most frequently used ML algorithms. In Sect. 6, a case presentation with the application of machine learning algorithms for diabetes detection is analyzed, with special reference to methodology, data collection, data preprocessing, Exploratory Data Analysis, [7] data visualization [8], evaluation metrics, conclusions, challenges, and future work. Section 7 presents big data software platforms and compares Spark vs Hadoop [10] as the leading open-source big data infrastructure software packages used to store and process large datasets. In the eighth Sect. 8, a review of the privacy and security of big data is given [11]. Section 9 covers a discussion that implies the implementation of a sophisticated healthcare system in the context of big medical data. In the last, and Sect. 10, future risks and concluding observations are highlighted. Lower level headings remain unnumbered; they are formatted as run-in headings.

2 Medical Big Data

In the medical industry, large amounts of data called Medical big data are generated, which are characterized by a high degree of irregularity, heterogeneity, dimensionality and untimeliness, due to which the true value of this data is not fully utilized. The effective use of medical big data implies the use of different strategies and techniques to exploit their true value [12]. The "10 Vs" concept is a framework that refers to a set of characteristics related to the management and analysis of medical big data and includes: volume, speed, variety, veracity, value, variability, vulnerability, volatility, validity, and visualization. The sheer volume of medical data is occurring due to the continuous, real-time generation at breakneck speed of voluminous electronic health records. The diversity of medical data occurs as a result of the different formats in which they appear, which imposes the need for their integration and comprehensive analysis. The truth of medical data implies their reliability and accuracy by minimizing possible errors and inconsistencies, extracting meaningful values for better patient outcomes, personalized treatments, disease prediction and better health system efficiency. Data variability occurs due to differences in patient populations, technical variations, healthcare practices, methods, and data collection sources. To deal with variability, advanced analytical methods are used to reduce variability. The vulnerability of medical data occurs due to its exposure to potential attacks and abuses due to its sensitive and personal nature. The instability of medical data occurs due to the rapid development of medical knowledge and practices, changing health status of patients, application of different measures and protocols, application of new methods and technologies, human error, etc. Coping with the instability of medical data is achieved by applying well-designed algorithms and analytical methods. The validity of medical data refers to the degree of accuracy of the information they contain, i.e. it is a measure of the accuracy of the aspects or phenomena to be measured, whereby systematic and random errors should be minimized. Visualization makes it possible to facilitate the understanding and interpretation of complex medical data in a practical way. Visual representations highlight inferences and anomalies that may not be apparent in the raw data. Specialized tools are used for visualization of medical data, Python libraries Matplotlib, Seaborn, Plotly, R libraries, Tableau, Power BI and others. Incorporating the "10 Vs" into the analysis and management of medical big data can lead to improved patient care, advances in medical research, and a better understanding of healthcare practices.

P4P [13] is a concept that focuses on progress in the field of medicine by integrating various aspects to improve patient care. The Predictive component refers to patient records, using genomic information and environmental factors to develop algorithms that can recognize disease risks before physical symptoms appear. The preventive component implies the prevention of diseases that are related to lifestyle, as well as screening for early detection of genetically predisposed conditions, healthy habits, regular examinations and tests. The Personalized component implies an individual approach to treatment, personalized treatments and therapies, which enables the reduction of potential risks. The participatory component allows patients to actively participate in decisions about their own health care and greater awareness of all health treatments, which can lead to better cooperation and results. The P4P concept, with the help of technologies such as genomics, big data analysis, and artificial intelligence, integrates the four P components, enabling efficient medical care.

The process of optimization and integration of Medical big data includes data preparation, cleaning and preparation of data in order to ensure their consistency. Mapping and transformation of data is carried out with the help of artificial intelligence algorithms in order to integrate data coming from different sources, as well as their appropriate storage. Algorithms can uncover complicated patterns and correlations that may not be obvious. Federated learning [14] is a new trend in artificial intelligence that allows training models with data that is distributed in different locations without violating privacy and security. Natural language processing (NLP) techniques analyze unstructured textual data from medical records and clinical records to extract meaningful insights. Image analysis with the help of artificial intelligence algorithms allow the detection of anomalies, tumors or other abnormalities with high accuracy. Big medical data requires scalable and high-performance systems that can process massive amounts of information in real time. In order to protect information and data, standardization in the medical industry and harmonized standards are needed.

Electronic health records (EHRs) [15] are real-time, patient-centered, digital versions of patient records that make information secure and available in real-time to authorized users. EHRs contain data about the individual's health status, including demographic data, diagnoses, therapy, results of performed analyzes and examinations, radiographic recordings, etc. Although the term "big data" is mentioned, the data on certain diseases, history of diseases, anamnesis, genetic data, previously performed medical treatments are quite poor, especially when it comes to rare diseases. EHRs are owned by healthcare facilities where the patient is treated and it is almost impossible to integrate these data, when it comes to public and private healthcare, and especially when it comes to different countries. A Personal Health Record (PHR) [16] is a digital system that allows individuals to track, store and manage their health data. A PHR can be in the form of a web-based platform, a mobile application, or a digital file. Creating a PHR personal health record owned and securely managed by the patient that gives them control and access to their personal health information is a solution for providing healthcare and evidence-based medicine.

Social influence is apparently channeled through interactions with friends rather than in a professional context, indicating that a patient may not fully disclose his condition to a physician, but may choose to communicate his feelings to his close friends on social networks. Social Networks Analysis – SNA uses networks and techniques for mining the content of social networks. "Social contagion" changes behaviors such as the decision to seek treatment, adoption and adherence to medical recommendations. Social networks can be integrated into clinical decision support systems, but there may be resistance within healthcare institutions because data from social networks is not sufficiently validated to be incorporated with other standards being tested.

Artificial intelligence uses large data sets coming from various sources, including electronic health records (EHRs), IoMT wearables and sensors, social network data, genomic data, etc. Presented are some of the most commonly used public datasets that are an essential starting point for scientists in the process of developing and training machine learning models. Medical Information Mart for Intensive Care III (MIMIC-III) [18] is a publicly available database used to develop predictive models, consists of de-identified electronic health records, contains a wide range of clinical data, including demographics, diagnoses, medicines, laboratory results, etc.

The Surveillance, Epidemiology, and End Results Program (SEER) [19] is a database in the United States that collects and publishes data on cancer incidence and survival. The SEER database includes demographic and clinical data for each patient compiled from a network of cancer registries. SEER provides a software tool called SEER*Stat that allows users to access, analyze, and visualize SEER data.

ImageNet [20] is a large-scale database created in 2009 by researchers at Stanford University that contains numerous labeled photographs spanning many categories, making it one of the most complete and widely used image datasets in the discipline of synthetic intelligence. ImageNet has played a significant role in advancing the field of computer vision, particularly with the rise of deep convolutional neural networks (CNNs).

UK Biobank [21] founded in 2006 is a large biomedical database and research initiative that aims to improve understanding of the causes of disease and promote advances in healthcare. UK Biobank stores a vast amount of health-related data making it a valuable resource for researchers studying a wide range of medical conditions and health factors.

3   Analytics in the Field of Large Amounts of Medical Data

Data analytics involves the development and application of algorithms for the analysis of various complex sets of big data. Although big data analytics promises a potential solution to a diverse range of problems, there remain many technical, computational, and statistical challenges that must be addressed to fully explore its potential.

Data mining [22] is a multidisciplinary field that brings together database technology, statistics, ML, and pattern recognition to predict disease, assess the risk of potential diseases, and help clinicians make the right decisions. Data mining, also known as knowledge discovery in databases, refers to the process of extracting potentially useful information and knowledge hidden in a large amount of incomplete, "noisy", vague and random, practical and application data [23]. The process of data mining shows a series of research concepts and is divided into several steps:

- Selection of a database according to the objectives of the research;
- Data extraction and integration, including retrieving the necessary data and combining data from multiple sources;
- Cleaning and transforming data, including removing incorrect data, filling in missing values, generating new variables, converting data format and ensuring data consistency;
- Data mining involves the extraction of implicit relational patterns through traditional statistics or machine learning;
- Creation and evaluation of models, which focuses on the validity parameters of the metrics;
- Evaluation of the results, which includes the interpretation of the extracted model in the form of knowledge available to the public.

Data mining technology does not aim to replace traditional statistical analysis techniques, but seeks to extend statistical analysis methodologies. Algorithms for mining large clinical data play a key role in advancing medical research. Clinical researchers have multiple data mining algorithms at their disposal, depending on the specific problem, data characteristics, interpretability requirements, and desired outcomes in the health care domain.

4 Machine Learning in the Field of Large Amounts of Medical Data

Machine learning is defined as "the ability of a machine to imitate intelligent human behavior" [4]. Machine learning is a set of algorithms and statistical models that computers use to perform a specific task without the use of explicit instructions. The goal of machine learning is to learn from the data collected for a specific task in order to maximize the machine's performance. Artificial intelligence leads to intelligence or wisdom, while machine learning leads to knowledge. Artificial intelligence aims to get the most appropriate solution, while machine learning considers only one answer, whether it is optimal or not. Machine learning algorithms build a scientific model that mainly relies on sample data that provide training data [24]. The model is designed to make predictions or make decisions without being explicitly programmed to perform the task. The intersection of different disciplines, especially mathematics, statistics and computer science, is an important essence of data science needed to implement different machine learning models. Machine learning can be categorized into several types based on the learning process, the nature of the data, and the desired results.

Supervised Learning [25] is a type of machine learning that is based on training the model with labeled data sets, which indicate to the system the relationship between input and output data. Supervised Learning can predict or classify data based on data from the training set, where the correct solutions or classifications are already known. Supervised Learning is based on training, so the model is trained until it discovers the underlying patterns and relationships between the input and output data. With Supervised Learning, the database is moved to Training Set and Test Set. Various metrics such as accuracy, precision, response, and others are used to measure how well a model predicts new data. Supervised Learning is used to solve problems related to regression and classification. Regression algorithms predict a continuous value of an output variable based on input characteristics. Models are evaluated using the usual metrics: mean squared error (MSE), root mean squared error (RMSE), mean absolute error (MAE) and R-squared (coefficient of determination). Common types of regression algorithms are: Linear Regression, Polynomial Regression, Ridge and Lasso Regression. Classification algorithms are used in machine learning to predict the classes or categories of objects or instances based on their characteristics or attributes. These algorithms are often used in tasks such as spam detection, medical diagnosis, image categorization and many others. Commonly used classification algorithms are: Logistic Regression, Decision Trees, Random Forest, Naive Bayes Classifier, Support Vector Machines, K-Nearest Neighbors – KNN, Neural Networks, Deep Neural Networks, Gradient Boosting.

Unlike Supervised Learning, where an algorithm works with labeled data to make predictions, Unsupervised Learning [26] focuses on discovering patterns, relationships, and structures within the data without explicit guidance. Clustering algorithms are fundamental techniques in unsupervised machine learning that group similar data based on certain characteristics or patterns that play a key role in understanding the inherent structure in the data, without looking for labeled samples. Clustering and dimensionality reduction are common tasks in unsupervised learning. Clustering algorithms divide the database into different clusters so that the data within a cluster are similar to each other. The most commonly used algorithms are:

K-Means Clustering, Hierarchical Clustering, DBSCAN (Density-Based Spatial Clustering of Applications with Noise, Gaussian Mixture Models (GMM), Principal Component Analysis (PCA), t-Distributed Stochastic Neighbor Embedding (t-SNE), Autoencoders, Self- Organizing Maps (SOM), Latent Dirichlet Allocation (LDA), Non-negative Matrix Factorization (NMF).

Semi-supervised learning [27] is a combination of supervised and unsupervised learning. The algorithm is trained on a dataset that includes both labeled and unlabeled data, so it uses the labeled data to apply knowledge to the unlabeled data. The most commonly used algorithms are: Semi-Supervised Support Vector Machines (S3VM), Co-Training, Self-Training, Label Propagation, Graph-Based Methods, Expectation-Maximization (EM) for Gaussian Mixture Models, Entropy Minimization, Transfer Learning, Pseudo- Labeling.

Deep learning [28] is a branch of machine learning that aims to develop a model that matches the level of the human brain in solving complex real-world problems using artificial neural networks and simulation learning. The human brain consists of a group of neurons, and deep learning attempts to reach the level of these neural connections by simulating the human brain through artificial neural network techniques. Deep learning technology is capable of recognizing images, communicating and translating from one language to another. Deep learning requires the use of high-level cognitive skills such as analysis and synthesis and enhances the understanding and application of lifelong learning. It is characterized by the speed of learning, because it has the ability to learn from large amounts of data, which humans cannot process. The most commonly used algorithms are Feedforward Neural Networks (FNN), Convolutional Neural Networks (CNN), Recurrent Neural Networks (RNN), Long Short-Term Memory (LSTM) Networks, Gated Recurrent Units (GRU), Autoencoders, Generative Adversarial Networks (GAN), Transformer, Deep Reinforcement Learning, Transfer Learning [29–31].

5 Comparative Analysis Among ML Algorithms

The analysis of medical data is performed using various algorithms and machine learning techniques. Each algorithm has its own strengths and weaknesses, and their correct application is vital for making informed decisions in real-world predictive modeling scenarios [32–34]. Table 1 shows a classification of the most commonly used ML algorithms in terms of application in healthcare, strengths, limitations, and selection of the prediction model.

6 Application of Machine Learning Algorithms for Diabetes Detection

Machine learning plays a significant role in improving diagnostic processes and risk prediction, personalized therapies, review of medical images, etc. Choosing the right algorithm in a real-world project involves careful consideration of various factors related to the data, problem, and research objectives.

6.1 Methodology

In this chapter a practical analysis of the most used ML algorithms for detecting the presence of diabetes is presented. Diabetes is a chronic disease that occurs when the immune system of the human body cannot function properly due to a change in the level of sugar in the body. Diabetes occurs due to genetic and environmental factors, and if it is not detected in time, it contributes to the development of other chronic diseases. There are several risk factors attributed to diabetes: overweight, unhealthy diet, physical inactivity and smoking [35]. The purpose of this study is to determine the risk of diabetes in patients based on data analysis. This comprehensive review will not only provide insight into diabetes risk patterns and trends, but will also create a solid foundation for further research. Specifically, research can build on how these variables interact and influence the onset and progression of diabetes, which is a key knowledge to improve patient care and outcomes in this increasingly critical area of health care. Therefore, machine learning methods have been proposed to be used by doctors for the purpose of easier diagnosis of diabetes.

Methodology includes a series of procedures that take place in several stages. The first phase involves gathering relevant medical data, including patient demographics, genetic information, medical history, and diagnostic features. Then, data cleaning is

Table 1. Comparative analysis of ML algorithms.

Heading level	application in healthcare	strengths	Limits	model selection
linear regression	Continuous prediction based on linear relationships	- simple, - interpretable - well suited	- limited to linear relationships - sensitive to outliers	- continuous target variable - models linear relationships between features and target
logistic regression	Binary classification using logistic function	- efficient, -interpretable - Suitable for probability-based predictions	- limited to linear relationships - sensitive to outliers	- in binary classification -need for probability-based predictions
decision trees	Predicting treatment outcomes and identifying risk factors	- efficient, -interpretable - Suitable for probability-based predictions	- overfitting, - lack of robustness, - inflexibility - insensitive to small changes in data	- clear explanation of decisions and interpretation
random forests	Disease prediction, medical image analysis and feature selection in health data	- efficient, -interpretable - Suitable for probability-based predictions	- Complexity can reduce interpretability, - the training time is longer, - Need for a large number of hyperparameters	- high accuracy and processing of large amounts of data, - if interpretation of decisions is critical, other models should be considered
KNN	Used for segmentation of patients based on similarities in health data, personalization of treatment	- Simple and efficient, - Perform well for identifying clusters, - scalable to large datasets	- Complexity can reduce interpretability, - the training time is longer, - Need for a large number of hyperparameters	- high accuracy and processing of large amounts of data, - if interpretation of decisions is critical, other models should be considered

(*continued*)

Table 1. (*continued*)

Heading level	application in healthcare	strengths	Limits	model selection
naive bass	Text classification, etc. two-dimensional data	- Simple and efficient, - Perform well for identifying clusters, - scalable to large datasets	- may require data cleansing or remote detection	-text classification, spam detection
XGBoost	- Diagnosis of diseases -Personalized treatment - Medical image analysis	- High accuracy: - Ability to handle various types of attributes - Works with incomplete data	- Assumption of independence of attributes - interpretation - Upgrading and tuning parameters:	For tasks where accuracy is critical, but other methods such as neural networks, random forests, etc. should be considered

performed in order to deal with missing values, analysis of outliers and data quality. To ensure uniform scaling, it is necessary to perform normalization and standardization of numerical characteristics and coding of categorical variables. Unnecessary features are removed in order to reduce dimensionality. Univariate or multivariate analysis techniques are used in order to examine the correlation between variables. The next stage is the selection of suitable algorithms for machine learning. The selected models are trained based on the training data. Models are evaluated using appropriate metrics for binary classification, such as accuracy, precision, recall, F1-score and area under the ROC curve (AUC-ROC), confusion matrix. It is good practice to visualize the results of the model to gain insight and interpretability. The continuous monitoring of the model is required as well as implementing mechanisms to update it when new data becomes available. We need to ensure that the use of patient data complies with ethical and legal regulations, such as HIPAA (in the United States) or GDPR (in Europe).

6.2 Data Collection

In our study, a database downloaded from Kaggle [36] was used, consisting of 100,000 rows containing 9 attributes, 8 of which are input attributes that include demographic data and clinical parameters and one target attribute that indicates the existence of diabetes risk. All 8 attributes cover a different aspect of the individual's health profile (gender, age, hypertension, heart disease, smoking_history, BMI, HbA1c_level, blood_glucose_level).

6.3 Data Preprocessing

Data cleaning is a key step in data processing, removing duplicates and resolving missing values and other inconsistencies or errors. The analysis and correction of null values is performed carefully, where the impact of missing values is considered, especially when their percentage is significant, and appropriate techniques are applied to replace them, providing optimal data. After a detailed examination of the data set, focusing specifically on facility types, a significant number of records (35,816) were observed where the smoking history status was marked as "No Information". Such status "No information" can be considered as a type of information, however, it is not possible to supplement or complete this information considering the fact that it covers more than 30% of observations, which means that smoking history is a significant predictor, improvement in data collection methodology is needed to ensure accurate recording of this information. Given the high percentage of "No information" values, it is recommended to remove this predictor from the analysis as it may introduce bias or inaccuracies into the analysis if missing value filling methods are used. We aim to maintain the integrity and quality of the data collected, ensuring reliable analysis and accurate interpretations. Table 2 presents the basic statistical parameters of the database.

Table 2. Basic statistical parameters

Heading level	Example	Font size and style	Heading level	Example	Font size and style	Heading level	Example
age	hypertension	heart_disease	bmi	HbA1c_level	blood_glucose_level	diabetes	target
count	99982.00	99982.00	99982.00	99982.00	99982.00	99982.00	99982.00
mean	41.89	0.07	0.04	27.32	5.53	138.06	0.09
std	22.52	0.26	0.19	6.64	1.07	40.71	0.28
min	0.08	0.00	0.00	10.01	3.50	80.00	0.00
25%	24.00	0.00	0.00	23.63	4.80	100.00	0.00
50%	43.00	0.00	0.00	27.32	5.80	140.00	0.00
75%	60.00	0.00	0.00	29.58	6.20	159.00	0.00
max	80.00	1.00	1.00	95.69	9.00	300.00	1.00

6.4 Exploratory Data Analysis

Exploratory Data Analysis (EDA) 0 is a fundamental step in diabetes risk prediction using machine learning. It involves a systematic approach to visually and statistically examine the database, uncovering essential insights and patterns. By uncovering the hidden insights in the data, EDA not only improves the accuracy of the predictive model, but also ensures the identification of clinically relevant factors, ultimately leading to more accurate and effective diabetes prediction models.

Table 3 shows the correlation of potential relationships between variables in order to highlight interactions and dependencies in the database. This analysis provides valuable insights into how variables potentially influence each other as well as their collective impact on diabetes risk. Visualization of the correlation matrix provides insight into which characteristics are positively or negatively correlated with each other and with the target variable.

Table 3. Correlation matrix

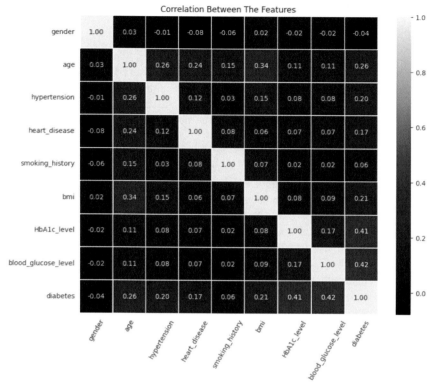

Table 4. Accuracy, precision, recall, F1-score in different ML models

model	accuracy	precision	recall	F1-score
Logistic Regression	96%	92%	81%	85%
Decision Tree	95%	84%	86%	85%
K-Nearest Neighbors	95%	91%	76%	82%
Random Forest	97%	96%	84%	89%
XGBoost	97%	100%	67%	80%

Data visualization plays a key role in diabetes prediction in the ML machine learning domain. Visualization enables healthcare professionals and data scientists to gain deeper insight into complex data sets, facilitating the identification of meaningful patterns and relationships between attributes. Visual displays, such as histograms, scatterplots, and ROC curves, enable a comprehensive understanding of data distributions, feature importance, and model performance metrics. These visualizations guide critical decisions in model selection, feature engineering, and threshold optimization, ultimately improving the accuracy and interpretability of ML diabetes prediction models, leading to more informed clinical decisions and improved patient care.

6.5 Evaluation Metrics

In predicting the risk of diabetes using machine learning, several evaluation metrics are crucial for assessing the performance of the prediction models 0, 0.

Accuracy measures the percentage of correctly classified cases among all cases and represents the ability to correctly identify true positive cases while minimizing false positives, which is critical for patient safety.

Accuracy in machine learning is the key metric in detecting diabetes or any other medical condition. Precision specifically measures the accuracy of the positive predictions made by the model, meaning it estimates how many of the predicted cases of diabetes are true positives. High accuracy indicates that when the model predicts that someone has diabetes, it is likely to be correct. In a medical context, achieving high accuracy is essential because misdiagnosing diabetes can have significant consequences for a patient's health and well-being.

Recall is known as sensitivity and measures the ability of the model to correctly identify individuals with diabetes from the total number of individuals who actually have diabetes. The high recall indicates that the model is effective in detecting the majority of true positive cases, ensuring that individuals with diabetes are less likely to be missed or falsely classified as non-diabetic. This is especially crucial in healthcare applications, since missing a true positive (diabetic patient) can lead to a delayed treatment and potential health risks. Therefore, achieving high recall is essential to ensure the model's ability to comprehensively identify individuals with diabetes in order to provide timely and appropriate medical interventions.

The F1-score is a critical metric for diabetes detection in machine learning because it provides a balanced evaluation of model performance, considering both precision and recall. In the context of diabetes detection, achieving a high F1-score is essential because it provides a harmonious compromise between minimizing false positives and false negatives. A high F1-score indicates that the model can make correct positive predictions (cases of diabetes) and also minimizes the risk of misclassifying individuals as diabetic when they are not or vice versa.

The ROC-AUC (Receiver Operating Characteristic - Area under the curve) score is a key evaluation metric in machine learning for binary classification tasks, including diabetes prediction. It quantifies the predictive model's ability to distinguish between positive and negative cases. The ROC-AUC score ranges between 0 and 1, with higher values indicating better discrimination. A score of 0.5 suggests that model performance is no better than a random chance, while a score of 1 indicates perfect discrimination. In the context of diabetes prediction, a high ROC-AUC score implies that the model effectively separates patients at risk of diabetes from those without, which is particularly valuable in healthcare as it indicates the ability of the model to make accurate and clinically relevant predictions.

The confusion matrix is a fundamental tool for evaluating the performance of machine learning models, especially in the context of diabetes prediction. It provides a detailed overview of model classification results by categorizing the predictions into four categories: true positives (correctly predicted cases of diabetes), true negatives (correctly predicted cases without diabetes), false positives (incorrectly predicted cases of diabetes), and false negatives (incorrectly predicted cases without diabetes). This matrix helps healthcare professionals and data scientists assess the model's ability to make accurate predictions, identify areas where it may be prone to error (such as false positives or false negatives), and fine-tune the model to achieve an appropriate balance between sensitivity and specificity.

ROCs (Receiver Operating Characteristic) are invaluable in the field of diabetes prediction using machine learning because they illustrate the trade-off between a model's true positive rate (sensitivity) and its false positive rate. By plotting the ROC curve, we can visually assess how well the predictive model discriminates between patients at risk of developing diabetes and those without risk of developing diabetes. In the context of diabetes prediction, ROC curves help determine the optimal threshold for clinical decision-making, providing an appropriate balance between correctly identifying patients at risk and minimizing false alarms, thereby improving the clinical utility of the model.

6.6   Results Comparisons

Evaluation Scores of ML Models are presented in Table 4. Their performance in terms of accuracy, precision, recall, F1-score was evaluated in order to build an efficient and accurate model for predicting the risk of diabetes.

Based on Table 4 we can make an evaluation of the five ML algorithms considered for predicting the risk of heart diseases.

In terms of accuracy both Random Forest and XGBoost achieved the highest accuracy of 97%. Logistic Regression, Decision Tree, and K-Nearest Neighbors also performed well, with 96% and 95% accuracy, respectively.

In terms of accuracy XGBoost achieved the highest accuracy of 100%, indicating that it made very accurate positive predictions. Random Forest also performed extremely well with 96% accuracy. Logistic regression and K-nearest neighbors had respectable accuracy scores of 92% and 91%, respectively. Decision Tree had the lowest accuracy at 84%.

In terms of recall Decision Tree had the highest recall of 86%, indicating its ability to identify a high percentage of relevant cases. K-Nearest Neighbors had a decent recall of 76%, while Logistic Regression and Random Forest scored 81% and 84%, respectively. XGBoost had the lowest recall of 67%.

In terms of F1-score Random Forest achieved the highest F1-score with 89%, showing balanced performance in terms of precision and recall. Logistic regression and decision tree had F1-scores of 85%. K-Nearest Neighbors had an F1-score of 82%. XGBoost had the lowest F1-score of 80%, indicating a trade-off between precision and recall.

Random Forest and XGBoost showed the highest accuracy and precision among the models, making them suitable for tasks where accuracy and precision are critical. Decision Tree, despite its slightly lower precision, achieved the highest recall and competitive F1 score, making it a balanced choice. K-Nearest neighbors provided a good balance between precision and recall. Choosing the best model depends on your specific use case and priorities, taking into account factors such as precision, recall, and the trade-offs between them. The choice of the best model ultimately depends on the specific application requirements and the relative importance of these metrics in the context of diabetes prediction.

6.7 Conclusion, Challenges, Future Work of Application of Machine Learning Algorithms for Diabetes Detection

The goal of the dataset analysis was to build a diabetes prediction model and compare the performance of different machine learning algorithms. The latest algorithms use advanced machine learning approaches and openly available patient data to predict patients' risk of developing diabetes. The methodology must be constantly improved and developed, so that other clinical parameters can be included.

Random Forest can be computationally expensive, especially for large data sets, and can require significant memory resources. Tuning hyperparameters can be challenging, prone to overload if not properly optimized. Random Forest is known for its ability to handle high-dimensional data and capture complex relationships. Future challenges related to Random Forest are addressed by developing more efficient and scalable implementations to deal with big data, exploring new hyperparameter optimization techniques. Exploring hybrid models that combine Random Forest with other algorithms to take advantage of each.

XGBoost can be computationally intensive and requires careful tuning of the hyperparameter. Its complexity may limit its interpretability. XGBoost achieves high performance due to its gradient boost framework. It can handle missing data, making it a powerful tool for predicting diabetes. Future challenges related to XGBoost include developing more efficient and parallelizable implementations to improve scalability. Research on advanced regularization techniques aimed at mitigating overload, improving model generalization, improving model interpretability through advanced visualization and Explainable artificial intelligence (XAI).

Deep learning algorithms such as Convolutional Neural Networks (CNN) and Recurrent Neural Networks (RNN) require large amounts of labeled data, significant computational resources, and extensive hyperparameter tuning. Interpreting patterns can be challenging in complex deep-learning architectures. Research on the application of deep learning models for diabetes prediction, considering their ability to capture complex relationships is focused on developing strategies for model explanation and interpretability, incorporating advanced data augmentation and synthesis methods to address missingness issues of data.

7   Big Data Software Platforms

Big data platforms and tools form the backbone of modern data-driven systems, enable companies to efficiently store, process and extract insights from massive and diverse data sets. Platforms such as Apache Hadoop and Apache Spark enable distributed storage and parallel processing. NoSQL databases like HBase and MongoDB 0 provide scalable and flexible storage solutions, and cloud services from providers like Amazon Web Services

(AWS) and Google Cloud offer managed big data services for seamless scalability and reduced infrastructure complexity. With their collective capabilities, big data platforms and tools play a key role in turning raw data into valuable insights and improved decision-making and innovation.

Apache Hadoop 0 is a software framework for distributed storage and processing of large amounts of data. Hadoop is an open source software package that uses the MapReduce programming model, which offers developers the flexibility to perform operations on large amounts of data. It is used when there is a need for large amounts of data to analyze structured and unstructured data. The Hadoop configuration uses a multi-server computing architecture that makes it economical to scale and support a huge data set. Hadoop Distributed File System (HDFS) 0, 0 is a distributed file system that stores data on many computers in a cluster. This provides high aggregate bandwidth and fault tolerance through data replication. Hadoop YARN (Yet Another Resource Negotiator) is a platform for managing computing resources in clusters. There is a rich ecosystem with various additional software packages and tools: Apache Pig, Apache Hive, Apache HBase, Apache Spark, Apache ZooKeeper and others. Hadoop is used in a number of applications and scenarios, including data analysis, machine learning, web browser processing, and many other areas where large amounts of data need to be worked with.

Apache Spark 0 is an open source distributed processing system commonly used to process large amounts of data, significantly faster than traditional batch processing systems. Its architecture supports a variety of data processing tasks, including batch processing, interactive queries, real-time stream processing, and machine learning. By providing high-level APIs in languages like Scala, Java, Python, and R, Spark enables developers and data scientists to easily build complex applications. Spark includes libraries for machine learning (MLib), graph processing (GraphX), and stream processing (Structured Streaming), further enhancing its versatility. With its ability to manage diverse workloads and simplify complex data processing tasks, Apache Spark 0 has become a cornerstone in the field of big data, enabling companies to extract valuable insights from massive data sets and drive informed decision making from data. Spark comes with many more innovative features, which enable it to achieve much more than what is possible with MapReduce Hadoop.

7.1 Spark vs Hadoop

Hadoop and Spark are software platforms, both developed by the Apache Software Foundation, that are widely used open source frameworks for big data architectures. Table 5 shows a comparison between Spark vs Hadoop.

Choosing the right platform depends on the nature of the project. Hadoop is chosen if there are many batch processing tasks, especially if there are significant amounts of records to be stored and processed. Spark is chosen if additional interactive and iterative or real-time processing is needed. Apache Spark, on the other hand, is recommended when there is a need for interactive, iterative or real-time processing. Spark provides a commitment to performance and is capable of working with massive amounts of data, making it an excellent choice for tasks such as machine learning, graph algorithms, and real-time data processing, such as processing data streams. The choice between Hadoop

Table 5. Comparison between Spark and Hadoop

Aspects for comparison	Spark	Hadoop
Architecture	It uses an in-memory processing model, enabling faster data processing by keeping intermediate results in memory. It supports batch processing, interactive queries, real-time stream processing and machine learning	It relies on a disk-based processing model, where intermediate results are written to disk after each processing stage. Mainly known for its MapReduce batch processing framework
Performance	It offers significantly faster processing times compared to Hadoop's MapReduce due to its in-memory computation, making it suitable for iterative algorithms and interactive data analysis	MapReduce can be slower due to the need to write intermediate results to disk, leading to higher latency for complex tasks
Ease of use:	It provides high-level APIs in multiple languages (Scala, Java, Python, R), making it more accessible to developers and data scientists. It offers interactive shells for quick experimentation	It provides high-level APIs in multiple languages (Scala, Java, Python, R), making it more accessible to developers and data scientists. It offers interactive shells for quick experimentation
Data Processing Paradigm:	It supports a wider range of processing paradigms, including batch, interactive, real-time stream processing, machine learning, and graph processing	Primarily known for batch processing using the MapReduce paradigm
Resource Management:	It supports multiple cluster managers, including Apache Mesos, Hadoop YARN, and Kubernetes	Originally integrated with Hadoop's YARN (Yet Another Resource Negotiator) for resource management and job scheduling
Ease of installation:	It is generally considered easier to install and set up, especially when used with Hadoop clusters	Setup can be more complex, especially for multi-node clusters

(continued)

Table 5. (*continued*)

Aspects for comparison	Spark	Hadoop
Ecosystem and Libraries:	It offers libraries like MLlib for machine learning, GraphX for graph processing, and Structured Streaming for real-time data processing	It provides a range of projects like Hive (SQL-like query), Pig (scripting for data analysis), HBase (NoSQL database) and more in its ecosystem
Use cases:	Well suited for iterative algorithms, machine learning, real-time stream processing, and situations where interactive analysis is important	It is traditionally used for batch processing of large amounts of data where low latency processing is not critical
Community and Adoption:	It has gained rapid adoption due to its performance advantages and versatile processing capabilities	It has a well-established community and is widely adopted in batch processing enterprises

and Spark depends on the specific needs and requirements of the project, as well as the type of data processing.

8 Privacy and Data Security

Data privacy and protection entails ensuring confidentiality, integrity and security by implementing appropriate measures that include: encryption, access controls, authentication and authorization, audit trails, cyber security measures, patient consent and transparency, ethical considerations, continuous monitoring and improvement.

The GDPR is considered as one of the most significant changes in data privacy regulation in the last twenty years. The GDPR was adopted by the European Parliament, the European Commission, and the Council of the European Union on April 27, 2016, and entered into force on May 25, 2018. HIPAA 0, 0 is a US law that was enacted in 1996, designed to protect sensitive medical information that is electronically transmitted and received. HIPAA reduces medical costs by allowing healthcare administrators to use electronic documentation and records.

The European Health Data Space EHDS builds on the General Data Protection Regulation (GDPR) 0, 0 and the NIS 2 Directive, consisting of rules, common standards and practices, infrastructures and a governance framework. Its goal is the increased digital access and control of electronic personal health data in the EU, providing a secure and efficient framework of health data for research, innovation, policy making and regulatory activities (secondary use of data). EHDS enables easy data sharing with healthcare professionals in and across member states, full control of citizens over their medical data, the ability to restrict access to others and the ability to obtain information about how their data is being used and for what purpose. The EHDS creates a strong legal framework for the use of health data.

9 Discussion

Artificial intelligence has brought great improvements in clinical data studies compared to traditional approaches. There are many limitations in applying machine learning to clinical data, due to its volume and difficulty to obtain. Although we are witnessing an explosion of healthcare data, there are few available open datasets for clinical studies. Most of the valuable datasets are owned by specific hospitals or research institutions. To train a reliable and efficient model, large training datasets are required. The small volume of samples from some rare diseases makes it difficult to use deep learning approaches. Clinical data comes in different formats, and combining data from different sources and the mechanism to merge them are both vital issues. Big data analysis mainly focuses on the perspectives of machine learning for personalized medicine, genomic data models and the application of data mining algorithms. Encouragingly, deep learning methodologies have improved predictive models, however, the interpretability of such models remains a challenge in the future. Computer vision algorithms can be trained to identify abnormalities in medical images that may indicate a disease. Healthcare robotics can be used to automate routine tasks, such as drug delivery and patient monitoring. It is necessary to choose a suitable big data platform that will have all the required libraries including machine learning libraries. The rapid development of healthcare applications raises privacy concerns, so blockchain technology could be a good solution. To increase the formality and standardization of data mining methods, a new programming language may need to be developed specifically for this purpose, as well as new methods capable of addressing unstructured data such as graphics, audio, and handwritten text.

10 Future Challenges

In the future, the healthcare system will need to use an increasing volume of big data with a higher dimension. Data analysis tasks and objectives will also have greater demands, including higher degrees of visualization, results with increased accuracy, and stronger real-time performance. In order to model an interpretable human-like computing system, it may be useful to join deep learning methods with medical ontologies, rule-based systems, and traditional machine learning solutions, and developing systems that can perform real-time analysis of continuous vital signs will help medical staff to promptly observe life-threatening pathological changes and provide appropriate treatment as early as possible.

In terms of application, the development of data management and disease screening systems for large-scale populations will help determine the best interventions and formulate supporting standards capable of benefiting both economics and personnel. The development of a scalable healthcare analytics application requires a fusion of different technologies whose selection requires careful scrutiny, as successful diagnosis and prevention of disease relies on incorporating as many data sources as possible. The first challenge is to choose from the big data platforms which will have all the necessary libraries including the machine learning libraries. The most difficult task in developing a stable solution is aggregating all the data from different sources. With advances in human genome research, personalized healthcare may now be a reality, however, matching genome sequences to medically relevant phenotypes is still a challenging topic.

The rapid development of health apps raises privacy concerns. Blockchain technology can be a good solution. Interoperability issues, technical issues, costs, security issues, privacy issues, lack of relevant policies and awareness of people's adoption/acceptance can be cited as future challenges. Due to the continuous growth of health care costs, it is necessary to strike a balance between costs and the quality of medical services. Much more work is needed to adequately protect the privacy of consumers using health apps ranging from educational campaigns to regulatory reform.

References

1. Aggarwal, K., et al.: Has the future started? The current growth of artificial intelligence, machine learning, and deep learning. Iraqi J. Comput. Sci. Math. **3**(1), 115–123 (2022)
2. Dimitrov, D.V.: Medical Internet of Things and big data in healthcare. Healthc. Inform. Res. **22**(3), 156–163 (2016)
3. Savoska, S., Ristevski, B., Trajkovik, V.: Personal health record data-driven integration of heterogeneous data. In: Data-Driven Approach for Bio-medical and Healthcare, pp. 1–21. Springer, Singapore (2022)
4. Nti, I.K., et al.: A mini-review of machine learning in big data analytics: applications, challenges, and prospects. Big Data Mining Analyt. **5**(2) (2022)
5. Cioffi, R., Travaglioni, M., Piscitelli, C., Petrillo, A., De Felice, F: Artificial intelligence and machine learning applications in smart production: progress, trends, and directions. Sustainability- MDPI **12**(492), 1–26 (2020)
6. Uddin, S., et al.: Comparing different supervised machine learning algorithms for disease prediction. BMC Med. Informat. Decis. Making **19**(1), 1–16 (2019)
7. Indrakumari, R., Poongodi, T., Jena, S.R.: Heart disease prediction using exploratory data analysis. Procedia Comput. Sci. **173**, 130–139 (2020)
8. Petrovski, G., Savoska, S., Ristevski, B., Bocevska, A., Jolevski, I., Blazheska-Tabakovska, N.: Visual Data Analysis for EU Public Sector Data using the app MyDataApp (2022)
9. Hicks, S.A., et al.: On evaluation metrics for medical applications of artificial intelligence. Sci. Rep. **12**(1), 5979 (2022)
10. Aziz, K., Zaidouni, D., Bellafkih, M.: Real-time data analysis using Spark and Hadoop. In: 2018 4th International Conference on Optimization and Applications (ICOA). IEEE (2018)
11. Hathaliya, J.J., Tanwar, S.: An exhaustive survey on security and privacy issues in Healthcare 4.0. Comput. Commun. **153**, 311–335 (2020)
12. Tai, M.: The Impact of. Artif. Intell. Human Soc. Bioethics **34**(4), 339–343 (2020)
13. Musich, S., et al.: Personalized preventive care reduces healthcare expenditures among Medicare Advantage beneficiaries. Am. J. Manag. Care **20**(8), 613–620 (2014)
14. Adnan, M., Kalra, S., Cresswell, J.C., Taylor, G.W., Tizhoosh, H.R.: Federated learning and differential privacy for medical image analysis. Sci. Rep. **12**, 1953 (2022)
15. Mohsen, F., et al. Artificial intelligence-based methods for fusion of electronic health records and imaging data. Sci. Rep. **12**(1), 17981 (2022)
16. Yousef, C.C., et al.: Perceived barriers and enablers of a personal health record from the healthcare provider perspective. Health Inf. J. **29**(1), 14604582231152190 (2023)
17. Yu, Z., et al.: Spatial correlations of land-use carbon emissions in the Yangtze River Delta region: a perspective from social network analysis. Ecol. Indicat. **142**, 109147 (2022)
18. Röösli, E., Bozkurt, S., Hernandez-Boussard, T.: Peeking into a black box, the fairness and generalizability of a MIMIC-III benchmarking model. Sci. Data **9**(1), 24 (2022)

19. Gupta, S., et al.: Prediction performance of deep learning for colon cancer survival prediction on SEER data. BioMed Res. Int. **2022**, 1–12 (2022)
20. Chawane, S.: Image based bee health classification. MS thesis. University of Twente (2022)
21. Conroy, M.C., et al.: UK Biobank: a globally important resource for cancer research. Br. J. Cancer **128**(4), 519–527 (2023)
22. Herland, M., Khoshgoftaar, T.M., Wald, R., Access, O.: A review of data mining using big data in health informatics. J Big Data **1**, 2 (2014). https://doi.org/10.1186/2196-1115-1-2
23. Kumar, S., Singh, M.: Big data analytics for healthcare industry: impact, applications, and tools. Big Data Min. Analyt. **2**(1), 48–57 (2019). https://doi.org/10.26599/BDMA.2018.902 0031
24. Hassanien, A.-E., Chang, K.-C., Mincong, T. (eds.): Advanced Machine Learning Technologies and Applications: Proceedings of AMLTA 2021, vol. 1339. Springer, Cham (2021)
25. Buczak, P., et al.: The machines take over: a comparison of various supervised learning approaches for automated scoring of divergent thinking tasks. J. Creative Behav. **57**(1), 17–36 (2023)
26. Chen, B., et al.: Mining tasks and task characteristics from electronic health record audit logs with unsupervised machine learning. J. Am. Med. Inf. Assoc. **28**(6), 1168–1177 (2021)
27. Solatidehkordi, Z., Zualkernan, I.: Survey on recent trends in medical image classification using semi-supervised learning. Appl. Sci. **12**(23), 12094 (2022)
28. Richards, B.A., Lillicrap, T.P., Beaudoin, P.: A deep learning framework for neuroscience. Nat. Neurosci. **22**(11), 1761–1770 (2019)
29. Lake, B.M., Ullman, T.D., Tenenbaum, J.B., Gershman, S.J.: Building Machines that Learn and Think Like People, vol. 40, no. E253, pp. 1–72. Cambridge University Press (2016)
30. Voulodimos, A., Doulamis, N., Doulamis, A., Protopapadakis, E: Deep learning for computer vision: a brief review. Comput. Intell. Neurosci. **2018**, 7068349 (2018)
31. Farhan, B.I., Jasim, A.D.: A survey of intrusion detection using deep learning in Internet of Things. Iraqi J. Comput. Sci. Math. **3**(1) (2022)
32. Ganie, S.M., Malik, M.B.: Comparative analysis of various supervised machine learning algorithms for the early prediction of type-II diabetes mellitus. Int. J. Med. Eng. Inf. **14**(6), 473–483 (2022)
33. Saxena, R., et al.: A Comprehensive review of various diabetic prediction models: a literature survey. J. Healthc. Eng. **2022**, 1–5 (2022)
34. Laila, U.E., et al.: An ensemble approach to predict early-stage diabetes risk using machine learning: an empirical study. Sensors **22**(14), 5247 (2022)
35. Chang, V., et al.: An assessment of machine learning models and algorithms for early prediction and diagnosis of diabetes using health indicators. Healthc. Analyt. **2**, 100118 (2022)
36. https://www.kaggle.com/datasets/iammustafatz/diabetes-prediction-dataset/data
37. Wang, G., et al.: Intelligent prediction of slope stability based on visual exploratory data analysis of 77 in situ cases. Int. J. Mining Sci. Technol. **33**(1), 47–59 (2023)
38. El-gezawy, M., Asmaa, M., Abdel-Kader, H., Ali, A.H.: A new XAI evaluation metric for classification. Int. J. Comput. Inf. **10**(3), 58–62 (2023)
39. Saha, S., et al.: MedTric: a clinically applicable metric for evaluation of multi-label computational diagnostic systems. PloS One **18**(8), e0283895 (2023)
40. Gui, Z., Paterson, K.G., Tang, T.: Security analysis of {MongoDB} queryable encryption. In: 32nd USENIX Security Symposium (USENIX Security 23) (2023)
41. Carvalho, I., Sá, F., Bernardino, J.: Performance evaluation of NoSQL document databases: couchbase, CouchDB, and MongoDB. Algorithms **16**(2), 78 (2023)
42. Zhang, J., Lin, M.: A comprehensive bibliometric analysis of Apache Hadoop from 2008 to 2020. Int. J. Intell. Comput. Cybernet. **16**(1), 99–120 (2023)

43. Laham, M.F., et al.: Performance Analysis of Apache Hadoop Using Hive on COVID19 Datasets
44. Chicco, D., Petrillo, U.F., Cattaneo, G.: Ten quick tips for bioinformatics analyses using an Apache Spark distributed computing environment. PLOS Comput. Biol. **19**(7), e1011272 (2023)
45. Luo, C., et al.: MapReduce accelerated attribute reduction based on neighborhood entropy with Apache Spark. Exp. Syst. Appl. **211**, 118554 (2023)
46. Shachar, C., et al.: HIPAA is a misunderstood and inadequate tool for protecting medical data. Nat. Med. 1–3 (2023)
47. Marks, M., Haupt, C.E.: AI chatbots, health privacy, and challenges to HIPAA compliance. JAMA **330**(4), 309 (2023)
48. Belen-Saglam, R., et al.: A systematic literature review of the tension between the GDPR and public blockchain systems. Blockchain: Res. Appl. **4**(2), 100129 (2023)
49. Johnson, G.A., Shriver, S.K., Goldberg, S.G.: Privacy and market concentration: intended and unintended consequences of the GDPR. Manag. Sci. **69**(10), 5695–5721 (2023)

Object Detection in Precision Viticulture Based on Uav Images and Artificial Intelligence

Milan Gavrilović[✉], Dušan Jovanović, and Miro Govedarica

Faculty of Technical Sciences, University of Novi Sad, Trg Dostiteja Obradovića 6, 21000 Novi Sad, Serbia

{milangavrilovic,dusanbuk,miro}@uns.ac.rs

Abstract. The system of precise viticulture is the basis for the improvement of conventional intensive viticulture production, while achieving yields of high quality and minimizing costs. Detection of vines in the vineyard is one of the key issues that can be solved with new technologies. Although the detection and the location of individual vine stems is of fundamental importance, there are no methods that can make this fully automatic. Various events such as diseases or mechanical damage led to the disappearance of plants over the years and the reduction of the initial number of vine stems per hectare. In order to accurately estimate the yield of the vineyard, it is necessary to identify and count the actual number of vines in order to estimate the yield based on them. A conventional method of such visual counting would be extremely time-consuming. This research aims to propose an automated model that can detect and locate living vine stems, using UAV imagery and artificial intelligence methods. The proposed model showed a high accuracy in the detection of vine stems. The advantage of the proposed model for the detection and identification of vines is that the results are obtained quite quickly and not much data is required (only RGB + NIR image of the vineyard from two periods without and with vegetation).

Keywords: remote sensing · YOLO · UAV · multispectral images · precision viticulture

1 Introduction

Precision viticulture represents the merging of the old concept of grape production with new technologies, and therefore enables winegrowers to make better decisions to optimize vineyard performance. The application of remote sensing and artificial intelligence in the precision viticulture sector is still not sufficiently developed, although there are many processes in viticulture that could be significantly improved using these technologies. In recent years, viticulture has been facing difficulties caused by the lack of qualified workers, which affects productivity, quality and timely harvest [1], and the tasks performed by hand are time-consuming and subject to the subjective influence of workers. These challenges motivate the development of new technologies, such as drones, and the use of artificial intelligence to ensure productivity, quality and economic competitiveness.

N. Filipović (Ed.): AAI 2023, LNNS 999, pp. 144–148, 2024.
https://doi.org/10.1007/978-3-031-60840-7_18

To accurately estimate the yield of the vineyard, it is necessary to identify and count the vines in the vineyard, in order to estimate the yield based on the actual number of vines. Given that the number of withered vine stems can exceed 20% [2], multiplying the average yield per culm with the theoretical number of vine stems can lead to a large overestimation. As visual counting is extremely time-consuming, it is necessary to develop an algorithm that can detect and locate living vine trees.

In most of the available literature [3–5], the authors mainly use 3D point clouds to determine the location of culverts, while a small number of works deal with the detection of vine stems based on aerial photographs. Different methods that can be found in the literature [1, 6, 7] can estimate the position of the vine stems, but using prior knowledge about the number of vine stems in the row and the distance between the trees. Unfortunately, vertical aerial photography often fails to identify the actual situation under the canopy, and in the absence of a plant, neighboring plants may spread their shoots and leaves to occupy adjacent free space [3]. Therefore, taking into account all the mentioned shortcomings, the authors of this paper propose a method that can detect and locate living vine trees, using only UAV (RGB + NIR) images and machine learning techniques.

2 Materials and Methods

2.1 Dataset Description

The experimental part of this research was conducted in a 1.3ha vineyard (45°11′16″N, 19°55′44″E) located in Sremski Karlovci (Fig. 1). This vineyard is part of the experimental field for viticulture of the Faculty of Agriculture of the University of Novi Sad. This vineyard was photographed during 2020 by a drone (DJI Phantom P4 v2.0) carrying a MicaSense RedEdge-M multispectral camera (B, G, R, Red-Edge, NIR) in two different periods of vine development (the first before the start vegetation, and the second during the vegetation period).

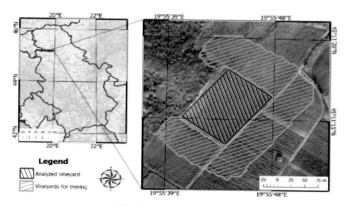

Fig. 1. Area of interest.

2.2 You Only Look Once (YOLO)

YOLO is neural networks-based algorithms used for object detection. This algorithm differs from other object detection algorithms in that it "observes" the image only once. The algorithm applies a single neural network to the entire image that divides that image into regions that provide bounding boxes and predicts probabilities for each region.

The design of the neural network was implemented through feedforward network, which means that during training and after it, information templates enter the network through the input layer, which then excites the hidden layers, so that the information ends up in the output layer. This means that the detection of objects on the entire image is performed in one run of the algorithm. Furthermore, the prediction of the probability of the object belonging to a certain class and the bounding boxes that specify the location of the object in the image is performed at the same time.

The features of the YOLO algorithm that make this class of algorithms widely used in the field of detection are speed, high accuracy and training capacity.

3   Results and Discussion

The detection of vines inside a vineyard using UAV images, as mentioned, is a challenge because in the later stages of vineyard development the vegetation is so dense that it is not possible to see the tree, and in the early stages when there is no vegetation the tree blends in with the soil and it is difficult to identify it on the images. This problem can be overcome by detecting the shadow cast by the vine tree in the first phase (Fig. 2), and then in the next phase localizing the vines.

Fig. 2. Vine before vegetation phase and its shadow.

Due to the lack of marked images with the position of vines and shadows, and in order to train the neural network, in the first step, vines are marked on vineyards that are in the immediate vicinity of the analyzed vineyard, and which are also affected when filming the analyzed vineyard with an unmanned aerial vehicle. The dataset created in this way and the neural network trained with it enable the application of knowledge transfer and the detection of vines in another vineyard without re-marking and creating training sets. By applying the previously trained model to a new data set, results are obtained very quickly with marked bounding boxes around each spikelet (Fig. 3). What can be seen is that a high degree of detection of individual trees is obtained, as well as that the model is not confused (columns are not identified as vine stems).

The next step is the localization of living vines, for which it is necessary to record the analyzed vineyard in the next phenological period, when the vineyard begins to

Fig. 3. The result of applying the YOLO algorithm for the analysed vineyard.

develop. The NDVI index is calculated with the help of the image created in this way (with vegetation), on the basis of which it can be concluded whether the culm has started to develop or not.

The next picture shows the method for localizing live vine stems. Localization of the vines is done by translating the boundary frames obtained first, with the identified shadow, until they cover the entire width of the row of vines (Fig. 4a). After that, using the NDVI index, the parts that belong to the vine rows are separated from the border frame (Fig. 4-b). The last step is to determine the coordinates of the live vine stems. The position of the culm is obtained by calculating the center point for the part of the row extracted by NDVI, which is located within the translated bounding box (Fig. 4-c).

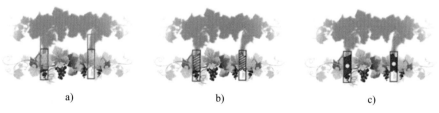

a)   b)   c)

Fig. 4. Determining the position of live bunches.

Based on the detection of vine stems using the previously described algorithm, 2368 live vine stems were detected, and field analysis determined the existence of 2569 live sedges. According to the results, the total accuracy of detection of roaches is 92.18% and it can be concluded that the proposed model is able to identify live vine stems with high accuracy.

4 Conclusions

The application of new technologies in viticulture enables automation and speeds up decision-making time. The advantage of the proposed algorithm for the detection and identification of grape vines is that the results are obtained quite quickly, without the need for 3D point clouds and a large number of other data.

In the further part of the research activities, it is necessary to further expand the proposed model with a part that would perform the detection and counting of withered vine stems in order to include all the key factors for precise viticulture. In addition, it is necessary to examine the possibility of using and applying the data obtained in this way during decision-making and management within the vineyard. It is also necessary to solve the problem caused by inter-row vegetation within the vineyard, which is reflected in the misinterpretation of the condition of the vines. In addition to UAV images, it is necessary to examine the possibility of using other geospatial data and, based on all these analyses and the obtained results, make appropriate conclusions about certain extensions and possibilities of further development of the proposed algorithm.

Acknowledgement. This research (paper) has been supported by the Ministry of Science, Technological Development and Innovation through project no. 451-03-47/2023-01/200156 "Innovative scientific and artistic research from the FTS (activity) domain".

References

1. Pérez-Zavala, R., Torres-Torriti, M., Cheein, F.A., Troni, G.: A pattern recognition strategy for visual grape bunch detection in vineyards. Comput. Electron. Agric. **151**, 136–149 (2018)
2. Matese, A., Di Gennaro, S.F.: Beyond the traditional NDVI index as a key factor to mainstream the use of UAV in precision viticulture. Sci. Rep. **11**, 2721 (2021)
3. Di Gennaro, S.F., Matese, A.: Evaluation of novel precision viticulture tool for canopy biomass estimation and missing plant detection based on 2.5D and 3D approaches using RGB images acquired by UAV platform. Plant Methods **16**, 91 (2020)
4. Jurado, J.M., Pádua, L., Feito, F.R., Sousa, J.J.: Automatic grapevine trunk detection on UAV-based point cloud. Remote Sens. **12**, 3043 (2020)
5. Comba, L., Biglia, A., Ricauda Aimonino, D., Gay, P.: Unsupervised detection of vineyards by 3D point-cloud UAV photogrammetry for precision agriculture. Comput. Electron. Agric. **155**, 84–95 (2018)
6. Aguiar, A.S., et al.: Grape bunch detection at different growth stages using deep learning quantized models. Agronomy **11**, 1890 (2021)
7. Vineyard yield estimation with smartphone imaging and AI. https://corbeauinnovation.com/vineyard-yield-estimation-with-smartphone-imaging-and-ai/. Accessed 20 Apr 2023

Material Classification of Underground Objects from GPR Recordings Using Deep Learning Approach

Daniel Štifanić[✉], Jelena Štifanić, Sandi Baressi Šegota, Nikola Anđelić, and Zlatan Car

Faculty of Engineering, University of Rijeka, Vukovarska 58, 51000 Rijeka, Croatia
{dstifanic,jmusulin,sbaressisegota,nandelic,car}@riteh.hr

Abstract. Exploration and detection of underground objectswithout excavation can be achieved by utilizing ground penetrating radar. Since such an approach is nondestructive, electromagnetic radiation has been used in order to accomplish sub-surface surveying. The correct interpretation of acquired ground penetrating radar data can be demanding, time-consuming and very challenging especially when the observed environment is noisy. However, with the assistance of artificial intelligence algorithms, such data can be processed and analyzed at high speed and with high accuracy. The aim of this research is to develop a deep learning model for the material classification of underground objects from ground penetrating radar recordings. Within the recordings, the pipes are usually visually represented as hyperbola-shaped features of different characteristics. Annotated by experts and preprocessed, ground penetrating radar recordings are used as input to the deep convolutional neural networks. In order to estimate the performances of the models, stratified 5-fold cross-validation is utilized along with the area under the ROC curve and confusion matrix. The results showed that the best performing model architecture (EfficientNetB3) can be used for pipe material classification from ground penetrating radar recordings with satisfactory results.

Keywords: ground penetrating radar · artificial intelligence · deep learning · convolutional neural network · stratified 5-fold cross-validation · classification

1 Introduction

Nowadays, it is essential for communal services such as water supply and energy sector to be equipped with modern technological solutions. By integrating such solutions into existing systems avoidable losses can be minimized, and thereby the overall efficiency can be increased [1]. Unfortunately, data that represent underground infrastructure that is stored one or more decades ago are usually incomplete and low-quality. However, by utilizing advanced devices, underground objects can be observed and explored from the ground surface without the need for excavation [2]. Such sub-surface surveying can be accomplished with electromagnetic radiation produced by ground penetrating radar (GPR). Using the aforementioned approach, the process of underground exploration is

N. Filipović (Ed.): AAI 2023, LNNS 999, pp. 149–158, 2024.
https://doi.org/10.1007/978-3-031-60840-7_19

simplified but the proper identification of underground objects can be very demanding and challenging which usually requires the support of a trained expert [3]. Data obtained by GPR can be processed in order to visually represent the observed underground as an image. In such cases, the width of the image corresponds to the distance traveled by the radar itself, and the height directly corresponds to the depth of electromagnetic wave penetration into the ground [2]. Images acquired in the aforementioned way can be used to train deep learning models based on convolutional neural networks [4].

In this research, underground objects of interest are pipes, more precisely different materials of pipes, such as metal and plastic. Within the GRP recordings, the pipes are usually visually represented as hyperbola-shaped features. Additionally, different pipe material will result in a feature with different characteristics within the GPR recording. Metal pipes usually produce strong reflections since the relative permittivity ε_r of the metal is infinity, thereby the features of such objects within the recording usually have visually stronger characteristics. However, the challenge arises when analyzing features produced by plastic pipes. In such cases, the electromagnetic wave can propagate through the pipe and the reflection feature mostly depends on the properties of a medium within the pipe. According to Baker et al. (2007), the ε_r of water is approximately 80 while the ε_r of air is 1 [5]. Since object reflections within the GPR recordings rely on numerous factors, the identification and classification of such is very time-consuming and labor-intensive work for trained experts. Additionally, the subjective factor of an expert can also have an impact on the interpretation of the recorded data. Therefore, in this research, the approach to process and analyze GPR data is by utilizing machine learning-based methods since a fully trained model can significantly reduce the time required for manual processing and analysis of such image data.

Deep convolutional neural networks (CNNs) basically revolutionized computer vision since they are designed to learn spatial hierarchies of features automatically and adaptively from given data [6]. Over the years, CNN-based models achieved cutting-edge results in many tasks including image classification, object detection, and semantic segmentation. In 2015, the ResNet architecture was presented by He et al. when it won the ILSVRC 2015 competition with an error of 3.6% on the ImageNet test set [7]. This research utilizes ResNet50 architecture in which, every 2-layer block is replaced (in the 34-layer network) with a 3-layer bottleneck block resulting in a total of 50 layers. The number of trainable parameters in such architecture is 23 888 771. Furthermore, in 2017, Chollet proposed a novel deep CNN architecture named Xception which is inspired by Inception. In such architecture, the Inception modules have been replaced with depth-wise separable convolutions with the aim of performance improvement [8]. The proposed architecture consists of 36 convolutional layers which are structured into 14 modules, all of which have so-called linear residual connections around them (not including the first and last module). In other words, the architecture can be described as a linear stack of depthwise separable convolution layers with residual connections. After the Inception architecture has been proven successful, Szegedy et al. in 2017 introduced InceptionRes-NetV2 architecture which combines Inception architecture with residual connections in order to significantly accelerate the training process [9]. InceptionResNetV2 is a 164-layer deep network that combines the strengths of the Inception modules and Residual connections to improve performance while maintaining computational efficiency. Tan and Le (2019) used neural architecture search in order to design a new network which is additionally scaled up to obtain a family of models, known as EfficientNets [10]. Such an approach results in achieving state-of-the-art 84.3% top-1 accuracy on ImageNet, while at the same time being smaller and faster compared to the best existing ConvNet (at the time). Moreover, the literature reveals that the number of research papers where machine learning-based methods are applied to GPR recordings has been increasing over the last few years. Ali et al. (2019) applied a multilayer perceptron classifier in order to process hyperbolic signature features which are extracted using statistical techniques. According to the obtained results, the proposed method for classifying different materials based on GPR images showed promising results [11]. A slightly different approach to identifying underground utility material from GPR imagery was introduced by El-Mahallawy and Hashim (2013). The authors used discrete cosine transform coefficients as features supplied to the support vector machine classifier [12]. Barkataki et al. (2022) presented an artificial neural network model for the automatic classification of buried objects. For the purpose of training and validation of the proposed model, the authors used synthetic data generated with gprMax software. According to the obtained results, the overall achieved accuracy was 95% [13]. Liu et al. (2020) in their paper proposed an automatic detection and localization method utilizing a deep learning approach along with migration. Experimental results showed that the detection accuracy of the proposed AI-based method is 90.9% [14]. It is known that environmental conditions, chloride contamination in concrete, and surface corrosion of rebars affect the amplitude of GPR signals, therefore Zatar et al. (2021) in their research utilized an AI-based method in order to analyze such effect. According to the obtained results, GPR reflection amplitudes can be estimated with a high level of accuracy with a coefficient of determination of 0.9958 [15]. Özkaya et al. (2021) proposed a combination of residual CNN and a bidirectional long short-term memory model for the GPR B-scan analysis. According to experimental results, the proposed model achieved an accuracy of 97.31%, 95.54%, and 93.15% in estimating GPR type, scanning frequency, and soil type, respectively [16].

The main contributions in this research are as follows:

- Investigating performances of various deep learning models in terms of material classification of underground objects.
- By utilizing fully trained machine learning-based models, the required time of well-trained experts for processing such data can be significantly reduced, and at the same time, objectivity and reproducibility can be improved.

2 Materials and Methods

The data used in this research consists of 2216 GPR recordings acquired in 17 different locations around the city of Rijeka, Croatia. Recordings are afterwards manually annotated by the experts which resulted in a total of 3794 patches representing the pipe feature. Patches can be divided into two classes, more precisely metal pipe class, and plastic pipe class. Moreover, the obtained dataset of patches is unbalanced with 2590 metal features and 1204 plastic features. Figure 1 shows the sample of annotated GPR recording where the bounding box outlined with a full line represents metal pipe feature while the boxes with dashed lines represent plastic pipe features.

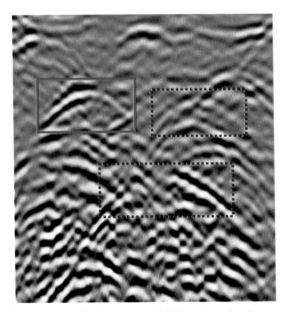

Fig. 1. GPR recording annotated by the experts; full line bounding box represents metal pipe feature and the dashed line bounding boxes represent plastic pipe features

Considering the related work from the previous chapter, in this research, ResNet50, Xception, InceptionResNetV2, and EfficientNet architectures are used to perform binary classification of pipe material from the aforementioned GPR recordings. In all cases at the top of the base architecture after the global average pooling layer, the dropout layer with the value of 0.2 and the output layer with two neurons and softmax activation function are added. Base layers of all architectures are pretrained on ImageNet while the added layers are initialized randomly. In the first stage, the newly added layers are trained using SGD, RMSprop, and Adam optimization algorithms with a learning rate of $1e-3$ and a learning rate decay of $1e-7$, while the base layers were frozen. In order to prevent overfitting, early stopping is utilized. In the second stage, the newly added layers were frozen while base layers are trained with the same optimization algorithms as in the first stage but with a lower value of learning rate ($1e-5$) and learning rate decay ($1e-9$) to ensure more stable training and convergence. The aforementioned training approach is used for all the architectures in question. Due to the high-imbalance of material classes, stratified 5-fold cross-validation is used to estimate the performance of artificial intelligence (AI) - based models. This way, the ratio between the classes in question is the same in each fold as it is in the full dataset.

To evaluate the performances of the aforementioned deep CNN architectures, the area under the ROC curve (AUC) metric is used. Visually, an ROC curve plots true positive rate (TPR) vs. false positive rate (FPR) at different classification thresholds [17]. Additionally, a confusion matrix is utilized to visually interpret the achieved results. This way, the relation between true and predicted values is observed in order to obtain classification performance of a two-class problem [18]. Confusion matrix is actually composed of true positive (TP), false positive (FP), true negative (TN) and false negative (FN) values. In the case of evaluating performances using the confusion matrix in binary classification problem, the threshold of 0.5 is used.

3   Results and Discussion

In this section, the experimental results achieved with ResNet50, Xception, Inception-ResNetV2, and EfficientNetB3 architectures are demonstrated and discussed. The models are trained in two stages with modified architecture configuration and hyperparameter combination, as described in the material and methods section. A comparison of best performances in terms of model architecture vs. optimization algorithm is shown in Fig. 2. Models are evaluated using AUC performance measure in combination with stratified 5-fold cross-validation. From Fig. 2. it is evident that in all four model architectures, the highest value of performance measure was achieved when the Adam optimization algorithm was used. Moreover, EfficientNetB3 architecture in the previously described configuration resulted in the highest mean AUC value of 0.93223 ± 0.01059 across 5-folds. Xception, ResNet50, and InceptionResNetV2 resulted in mean AUC values of 0.9316 ± 0.0144, 0.92536 ± 0.01249, and 0.91498 ± 0.01967, respectively.

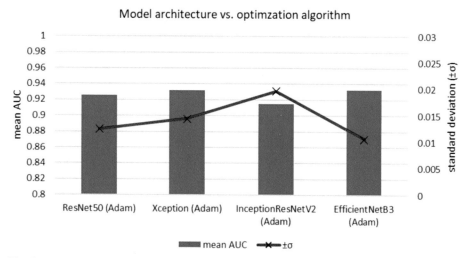

Fig. 2. Performances of ResNet50, Xception, InceptionResNetV2, and EfficientNetB3 architectures in terms of mean AUC with appurtenant standard deviation values.

Additionally, for the best performing architecture i.e., EfficientNetB3 (Adam), the confusion matrix was created in order to visualize the classification performance. In the case of this research, the confusion matrix elements are normalized over true conditions i.e., rows of the matrix. The confusion matrix of EfficientNetB3 trained with Adam optimization algorithm is shown in Fig. 3. The values of the matrix are averaged across 5-folds; therefore, each quadrant shows a mean value of an element along with the standard deviation. Since the matrix is normalized over true conditions, the standard

154   D. Štifanić et al.

deviation has the same value for each element in a row. Furthermore, it can be seen that in 90% ± 1.71 of cases the model successfully classified metal pipe. On the other hand, the model successfully classified 78% ± 3.13 of plastic pipes. According to the values of standard deviation, it can be concluded that features which represent plastic pipes are more challenging to classify correctly since in that case the model uncertainty is higher.

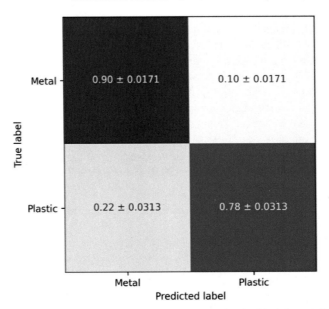

Fig. 3. Confusion matrix of EfficientNetB3 architecture trained with Adam optimization algorithm across 5-folds

A comparison of performances based on the lowest value of standard deviation across 5-folds is shown in Fig. 4. It can be seen that the ResNet50 model architecture trained with SGD achieved the lowest value of standard deviation (0.88027 ± 0.0082) but also a lower value of mean AUC. Analyzing performances of EfficientNetB3 model architecture trained with Adam and RMSprop optimization algorithms it can be concluded that when using RMSprop, the mean AUC is slightly lower (0.92165 ± 0.00834), and the value of standard deviation is also lower. In other words, the model estimations on validation sets over 5-fold cross-validation are more consistent for the cost of lower mean AUC value.

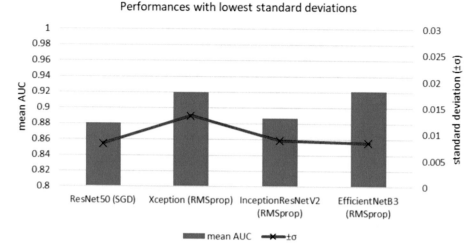

Fig. 4. Performances of ResNet50, Xception, InceptionResNetV2, and EfficientNetB3 architectures based on lowest standard deviation values

The confusion matrix of the EfficientNetB3 model architecture trained with RMSprop optimization algorithm is shown in Fig. 5. From the performances, it can be seen that the model in 87% ± 0.82 of cases correctly assigned a metal label to the input image. Moreover, in the 76% ± 2.11 of cases plastic pipe was correctly classified.

Since each model architecture is unique and complex in its own way, a different number of training epochs is needed to achieve convergence. Therefore, Fig. 6 shows the average number of epochs across 5-folds that was needed in each training stage before model AUC was no longer increasing.

The total average number of epochs needed (per fold) for training EfficientNetB3 with the Adam optimization algorithm was significantly higher (approx. 54 epochs) compared to other architectures. On the other hand, the same architecture resulted in the highest mean AUC value across 5-fold. Xception model architecture was completely trained after approx. 36 epochs per fold which is significantly faster than the training process of EfficientNetB3. Taking that into account, the mean AUC value of the Xception model architecture is only 0.00063 lower with a 0.00381 higher value of standard deviation compared to EfficientNetB3.

The presented results in this research show that the EfficientNetB3 architecture provides satisfactory results considering that the classification of pipe materials based on radar reflection is a very challenging task and many factors affect the outcome. Implementation of a fully trained classification model will be used in future work together with an object detection algorithm in order to provide additional information about the type of material the pipe is made of.

Confusion matrix - EfficientNetB3 (RMSprop)

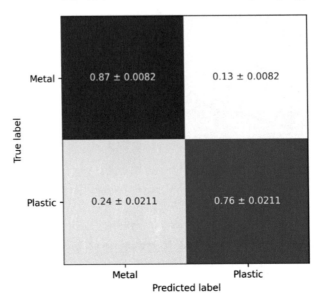

Fig. 5. Confusion matrix of EfficientNetB3 architecture trained with RMSprop optimization algorithm across 5-folds

Number of training epochs for different model architectures

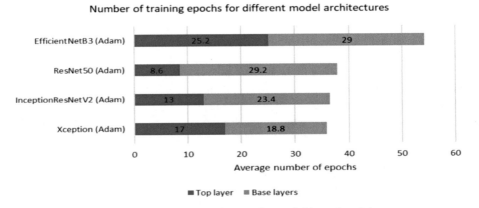

Fig. 6. Average number of training epochs needed in each training stage

4 Conclusions

According to the obtained performances of classification models, it can be concluded that deep CNNs have great potential in pipe material classification from GPR recordings. EfficientNetB3 model architecture adapted to the problem of this research along with the training process with frozen layers resulted in satisfactory results with a mean AUC value of 0.93223 ± 0.01059 in stratified 5-fold cross-validation. The unbalanced number of patches per class was a limitation of this research, therefore future work should use a more balanced dataset. The presented approach will be used in future work along with the object detection model in order to fully automate the process of detection, localization and classification of underground infrastructure.

Acknowledgment. This research was (partly) supported by the CEEPUS network CIII-HR-0108, the European Regional Development Fund under Grant KK.01.1.1.01.0009 (DATACROSS), the project CEKOM under Grant KK.01.2.2.03.0004, the Erasmus+ project WICT under Grant 2021-1-HR01-KA220-HED-000031177, and the University of Rijeka Scientific Grants uniri-mladi-technic-22-61 and uniri-tehnic-18-275-1447.

References

1. Krause, A., Perciavalle, P., Johnson, K., Owens, B., Frodl, D., Sarni, W., Foundry, W.: The digitization of water
2. Jol, H.M. (ed.) Ground penetrating radar theory and applications. Elsevier, 8 December 2008
3. Lu, Q., Pu, J., Liu, Z.: Feature extraction and automatic material classification of underground objects from ground penetrating radar data. J. Electr. Comput. Eng. 2014, 28 (2014)
4. Besaw, L.E., Stimac, P.J.: Deep convolutional neural networks for classifying GPR B-scans. In Detection and sensing of mines, explosive objects, and obscured targets XX 2015 May 21, vol. 9454, pp. 385–394). SPIE (2015)
5. Baker, G.S., Jordan, T.E., Pardy, J.: An introduction to ground penetrating radar (GPR)
6. Aloysius, N., Geetha, M.: A review on deep convolutional neural networks. In: 2017 International Conference on Communication and Signal Processing (ICCSP) 2017 Apr 6, pp. 0588–0592. IEEE (2017)
7. He, K., Zhang, X., Ren, S., Sun, J.: Deep residual learning for image recognition. In: Proceedings of the IEEE Conference on Computer Vision and Pattern Recognition, pp. 770–778 (2016)
8. Chollet, F.: Xception: Deep learning with depthwise separable convolutions. In: Proceedings of the IEEE Conference on Computer Vision and Pattern Recognition 2017, pp. 1251–1258 (2017)
9. Szegedy, C., Ioffe, S., Vanhoucke, V., Alemi, A.: Inception-v4, inception-resnet and the impact of residual connections on learning. In: Proceedings of the AAAI Conference on Artificial Intelligence 2017 Feb 12, vol. 31, No. 1 (2017)
10. Tan, M., Le, Q.: Efficientnet: rethinking model scaling for convolutional neural networks. In: International Conference on Machine Learning 2019 May 24, pp. 6105–6114. PMLR (2019)
11. Ali, H., Firdaus, A.A., Azalan, M.S., Kanafiah, S.N., Salman, S.H., Ahmad, M.R., Amran, T.S., Amin, M.S.: Classification of different materials for underground object using artificial neural network. In: IOP Conference Series: Materials Science and Engineering 2019 Nov 1, vol. 705, No. 1, p. 012013. IOP Publishing
12. El-Mahallawy, M.S., Hashim, M.: Material classification of underground utilities from GPR images using DCT-based SVM approach. IEEE Geosci. Remote Sens. Lett. **10**(6), 1542–1546 (2013)
13. Barkataki, N., Kalita, A.J., Sarma, U.: Automatic material classification of targets from GPR data using artificial neural networks. In: 2022 IEEE Silchar Subsection Conference (SILCON) 2022 Nov 4, pp. 1–5. IEEE (2022)
14. Liu, H., Lin, C., Cui, J., Fan, L., Xie, X., Spencer, B.F.: Detection and localization of rebar in concrete by deep learning using ground penetrating radar. Autom. Constr. **1**(118), 103279 (2020)

15. Zatar, W., Nguyen, T.T., Nguyen, H.: Predicting GPR signals from concrete structures using artificial intelligence-based method. Adv. Civil Eng. **1**(2021), 1–9 (2021)
16. Özkaya, U., Öztürk, Ş, Melgani, F., Seyfi, L.: Residual CNN+ Bi-LSTM model to analyze GPR B scan images. Autom. Constr.**1**(123), 103525 (2021)
17. Musulin, J., Štifanić, D., Zulijani, A., Ćabov, T., Dekanić, A., Car, Z.: An enhanced histopathology analysis: an ai-based system for multiclass grading of oral squamous cell carcinoma and segmenting of epithelial and stromal tissue. Cancers **13**(8), 1784 (2021)
18. Valero-Carreras, D., Alcaraz, J., Landete, M.: Comparing two SVM models through different metrics based on the confusion matrix. Comput. Oper. Res. **1**(152), 106131 (2023)

Genetic Programming Approach in Better Understanding of the Relationship Between the Number of Viable Cells and Concentration of O_2^-, NO_2^- and GSH Produced in Cancer Cells Treated with Pd(II) Complexes

Tamara M. Mladenovic[1($\boxtimes$)], Marko N. Živanović[1], Leo Benolić[2], Jelena N. Pavić[1], and Nenad Filipović[2,3]

[1] Department of Natural Sciences, Institute for Information Technologies Kragujevac, University of Kragujevac, Kragujevac, Republic of Serbia
tamaramladenovic09@gmail.com, jelena.grujic2@aiesec.net
[2] Bioengineering Research and Development Center (BioIRC), Kragujevac, Republic of Serbia
{leo.benolic, fica}@kg.ac.rs
[3] Faculty of Engineering, University of Kragujevac, Kragujevac, Republic of Serbia

Abstract. Genetic programming (GP) is a powerful tool for creating mathematical models and has been used in various fields to discover complex relationships in data. GP is a method that uses artificial intelligence to automatically create programs, based on the Darwinian principle of natural selection to create a population of better programs over many generations. Therefore, this study investigated the possibility of obtaining a function describing the data obtained from the research work conducted by Petrović et al., by applying genetic programming. Experimental results showed that Pd(II) complexes, labeled as Pd-1, Pd-3, Pd-5 and Pd-6, showed significant cytotoxic effects and extreme oxidative stress in cancer cell lines HCT-116 and MDA-MB-231. Specifically, the aim was to determine whether there is a concentration-dependent relationship between the number of viable cells for all experiments. To create the function, symbolic regression was used. R2 is a useful metric for evaluating the performance of the created function. Overall, the results demonstrate that GP is a powerful tool for creating mathematical models that accurately describe the relationship between concentration and the number of viable cells. In conclusion, it has been shown that GP can be used with high precision on such examples, with R2 ranging from 0.9747 to 0.99948. Such a method of calculating values could facilitate the selection of the required concentration for use in experiments.

Keywords: genetic programming · Pd(II) complexes · optimization · cytotoxic · oxidative stress

© The Author(s), under exclusive license to Springer Nature Switzerland AG 2024
N. Filipović (Ed.): AAI 2023, LNNS 999, pp. 159–169, 2024.
https://doi.org/10.1007/978-3-031-60840-7_20

1 Introduction

Since the discovery of cisplatin, there has been a tremendous increase in interest in searching for anticancer medicines derived from metal complexes. The anti-cancer drug cisplatin has been widely used for more than 30 years. Cisplatin's anticancer effect is dose-dependent and inhibits DNA synthesis. It has been successfully used in clinical trials to treat malignancies of the lung, head and neck, bladder, cervix, and ovaries (Achkar et al., 2018; Tchounwou et al., 2021). Cisplatin's antitumor action is acknowledged for its effectiveness in inducing programmed cell death (apoptosis) in cancer cells as a result of exposure (Aldossary, 2019; Brown et al., 2019). Unfortunately, resistance to cisplatin in some patients reduces the drug's therapeutic value. Additionally, cisplatin-containing medications have serious side effects such as neurotoxicity, cardiomyopathy, nephrotoxicity, liver damage, and ototoxicity (Sheth et al., 2017; Tchounwou et al., 2021). In response to these challenges, palladium(II) complexes have been found to be an effective substitute for cisplatin, offering benefits including fewer side effects and similar cytotoxicity against cancer cells to Pt-based medications (Feizi-Dehnayebi et al., 2021). It has been demonstrated that cancer cells exposed to Pd(II) complexes exhibited characteristics of programmed cell death (apoptosis) (Mbugua et al., 2020). Pd(II) complexes are particularly effective against breast cancer and gastrointestinal cancer when compared to platinum complexes (Feizi-Dehnayebi et al., 2021). Pd(II) complexes, labeled as Pd-1, Pd-3, Pd-5 and Pd-6, demonstrated considerable cytotoxic effects and high oxidative stress in cancer cell lines HCT-116 and MDA-MB-231 in a study conducted by Petrović et al. (Petrović et al., 2015). Therefore, this study served as a foundation for applying genetic programming to create a function that would describe the experimental data.

Nowadays, the use of machine learning techniques in delicate areas of daily life, such as natural sciences and medicine, is growing constantly (Filho et al., 2020). The goal of machine learning, commonly referred to as artificial intelligence, is to create computer systems that can adapt and learn from experience (Kavakiotis et al., 2017). Predicting the behavior of a dynamical system is the most important topic in all of science, especially in the natural sciences. These systems need an underlying mathematical model that can be used as a predictor (Quade et al., 2016). These mathematical models that depict solutions are frequently produced using the bio-inspired approach known as Genetic Programming (GP) (see Fig. 1). Applying cutting-edge data analysis algorithms that automatically learn from provided observations and generate models in the form of their own modeling languages, these algorithms deliver reliable estimations of observed dynamics (Filho et al., 2020; Quade et al., 2016).

The Darwinian idea of natural selection is the foundation of GP, a technique for employing artificial intelligence to automatically construct programs that develop over many generations into a population of superior programs (Koza, 1994). GP places a strong emphasis on the ability of the issue to calculate the quality of the potential solutions. If the problem satisfies this criterion, there should be a test to assess whether a solution actually resolves the issue, and it should be possible to classify two or more possible solutions in order of their quality (Eggermont, 2007). The most adaptable programs will survive and create offspring in a population during the evolution of programs, just like in nature. Every new individual must have a fitness value assigned to them in order to determine who will produce the offspring. This fitness value can be determined by

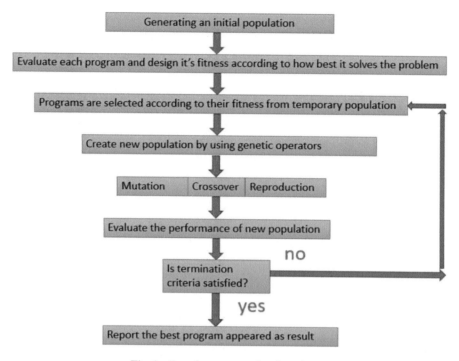

Fig. 1. Genetic programming flowchart.

running the program one or more times with a test set of inputs to observe how closely the algorithm produces the intended outputs (Langdon et al., 2013).

A functional programming language expresses the proposed solutions produced by genetic programming in syntax trees, also known as parse trees. Due to this, tree-based GP is another name for conventional GP (Zhang et al., 2021). Syntax trees encode how statements and expressions are nested to produce programs, with nodes representing operators and leaf nodes corresponding to operands (Liang et al., 2019). The syntax tree's population elements are terminals and functions, where functions can represent any mathematical operation (even simple arithmetic) and terminals can include arguments for functions (such as variables, numeric values, and logical contents) (Zhang et al., 2021).

In contrast to symbolic regression, conventional regression analysis only requires a limited number of function coefficients to be determined because the analyst has already specified the structure of the function (França, 2000). The process of developing a mathematical model from experimental data is known as empirical modeling. It is a real challenge to solve empirical modeling problems and find both the structure and appropriate numeric coefficients of a model. That is where GP is applied to solve this problem via symbolic regression. In order to predict a function's mathematical expression from its observed values, symbolic regression must typically be performed in two steps: first, it must predict the "skeleton" of the expression up to the selection of numerical constants. Then it must fit the constants by optimizing a non-convex loss function (Dabhi &

Chaudhary, 2015; Kamienny et al., 2022). The optimal model can be selected from a large class of candidates, where the candidates are explicit symbolic formulas (Dabhi & Chaudhary, 2015).

The procedure of optimization minimizes a certain measure of the distance between the response anticipated by the model (function) and the available data. GP is used to approach a symbolic regression when no specific model is provided as a starting point and must instead be found with all of its parameters (França, 2000). Therefore, using symbolic regression to create the function, the objective of this study was to determine whether there is a concentration-dependent relationship between the number of viable cells for all experiments.

2   Materials and Methods

2.1   Data from the *In Vitro* Experiment

Pd (II) complexes (Pd-1, Pd-3, Pd-5 and Pd-6) were diluted to operating concentrations (0, 0.1, 1, 10, 50, 100 μM), where 0 served as control (non-treated cells). Cancer cell lines MDA-MB-231 (breast cancer) and HCT-116 (colorectal cancer) were treated with different concentrations of Pd(II) complexes. In order to determine superoxide anion radical the NBT assay was done and for determination of NO_2^-, , the Griess assay was performed. The determination of reduced glutathione (GSH) was based on the redox reaction of intracellular GSH with Ellman's reagents.

Due to an increase in nitrite and superoxide anion radical generation, Pd(II) compounds have been shown to have prooxidant effects on the cancer cells HCT116 and MDA-MB-231. Increased cytotoxicity was caused by an increased production of reactive oxygen species (ROS) and reactive nitrogen species (RNS), particularly when Pd-1 and Pd-6 were used as treatments. The disruption of redox equilibrium caused by ROS/RNS is connected to the cell self-defense mechanism, which has an impact on the increased glutathione production.

Effects of Pd(II) complexes were shown after 24 h and 72 h of exposure and results were expressed in μM of O_2^-, NO_2^- and GSH from a standard curve established in each test, constituted of known molar concentrations of O_2^-, NO_2^- and GSH. The results were then calculated using statistical software programs (SPSS, ANOVA) and data were expressed as mean $\pm$ standard error (SE) (see Fig. 2) (Petrović et al., 2015). The obtained results served as a foundation for applying genetic programming to create a function that would describe this experimental data.

Superoxide anion radical $O_2^{\cdot-}$ (μM) after 24h of exposure

HCT-116

Complexes			
Pd-1	Pd-3	Pd-5	Pd-6
37.07±0.28	37.07±0.28	37.07±0.28	37.07±0.28
35.04±0.29	38.17±1.28	37.94±0.30	33.99±0.01*
39.32±0.48*	44.15±0.53*	38.60±0.12*	32.23±0.19*
37.88±1.23	41.79±0.50*	38.98±0.67*	32.16±0.44*
36.57±0.25	41.30±0.38*	43.62±0.56*	33.36±0.16*
39.19±0.72	41.05±0.95*	35.89±0.35	36.71±0.50
42.36±0.62*	54.85±0.47*	51.47±0.19*	46.96±0.52*

Nitrites, NO_2^- (μM) after 24 hours of exposure

HCT-116

Complexes			
Pd-1	Pd-3	Pd-5	Pd-6
43.21±0.34	43.21±0.34	43.21±0.34	43.21±0.34
51.92±1.17*	51.76±0.83*	47.24±2.34*	31.20±1.26*
45.46±2.08	35.96±0.64*	39.14±5.08	31.18±0.81*
47.17±1.15*	52.40±1.89*	62.07±1.55*	33.25±2.13*
27.90±0.46*	53.31±1.96*	47.86±1.59*	45.42±2.56
50.86±0.60*	53.93±0.83*	46.64±3.25	37.38±3.04*
61.74±0.51*	50.89±0.79*	51.95±1.01*	40.51±1.27

MDA-MB-231

Complexes			
Pd-1	Pd-3	Pd-5	Pd-6
31.02±0.19	31.02±0.19	31.02±0.19	31.02±0.19
29.46±1.34	31.15±0.21	32.51±0.74	34.81±0.21*
32.02±0.49	31.97±0.20	29.79±0.17	32.82±0.31*
32.00±1.61	32.15±0.76	34.46±1.08*	27.45±0.27*
35.70±0.41*	31.73±0.20	31.21±0.35	30.00±0.85
36.64±0.07*	34.65±0.05*	38.93±0.67*	30.24±0.57
37.43±1.37*	36.01±0.13*	47.63±0.56*	48.46±0.47*

MDA-MB-231

Complexes			
Pd-1	Pd-3	Pd-5	Pd-6
28.21±0.22	28.21±0.22	28.21±0.22	28.21±0.22
35.79±2.31*	57.98±2.98*	28.74±1.09	32.48±1.30*
38.89±0.57*	49.56±1.85*	35.20±3.68*	33.22±1.53*
57.43±0.49*	56.90±0.78*	30.72±1.98	24.22±0.98*
61.86±5.69*	52.17±3.67*	30.87±3.22*	31.20±2.47*
45.37±0.09*	56.72±3.08*	27.63±1.09	35.56±2.96*
52.31±1.73*	64.79±1.16*	32.99±0.51*	32.63±1.17*

Fig. 2. Segment of the scientific results used in Genetic Programming.

2.2 Symbolic Regression via Genetic Programming

To create the function, symbolic regression was used. This involved using genetic coding, selection, crossover, and mutation to create a tree structure, where nodes are functions and leaves are variables or constants (De Jong, 1988). A syntax tree form is used to define and display GP operations, crossover, and mutation. Crossover was carried out by choosing two "parent" trees, locating each one's root node, and then randomly choosing one of the two subtrees related to it. The chosen subtree was then eliminated at the end of the process. To make changes, any node from a given tree was simply chosen, and a random node was substituted in its place, ensuring that the new tree still represented the original formula.

In order to be imported into Python software for coding, the statistical data from the experiment were transferred in the form of a CSV (comma-separated values) file. However, due to the small dataset, the points were interpolated to increase the dataset size to maintain quality and divide it into a training and testing set in a 70% - 30% ratio. Interpolation and genetic programming were done in Python programming language (Wonjae Lee & Hak-Young Kim, 2005). Python uses the interpolation method to estimate unknown data points between two known data points. Interpolation is a Python preprocessing technique that is frequently used to fill in missing values in data frames or series.

The coefficient of determination R2 was used to evaluate the obtained solution's quality. R2 is a statistical measure representing the proportion of the variance for a dependent variable explained by an independent variable or variables in a regression model (Di Bucchianico, 2007). A high value of R2 (close to 1) indicates a better fit of the function to the data, while a low value of R2 (close to 0) indicates a poor fit. Therefore, R2 is a useful metric for evaluating the performance of the created function.

3   Results and Discussion

Examples of the resulting curves are shown in Fig. 3 and Fig. 4 for HCT-116 and MDA-MB-231 cell lines. The R2 values for these solutions are 0.99794 and 0.99802. From a total of 40 measurements, 32 solutions have an R2 value above 0.99, of which 20 values were obtained after 24 h of exposure and the other 20 after 72 h of exposure. One of the positive aspects of using GP is the ability to display the solution function as a syntax tree (see Fig. 5 and Fig. 6).

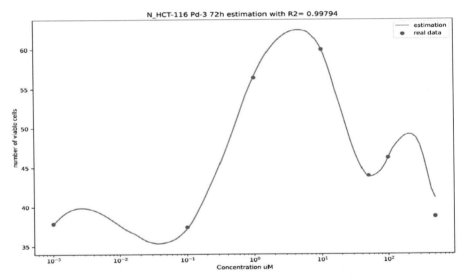

Fig. 3. The R2 value of produced nitrite of HCT-116 cell line treated with Pd-5 after 24 h.

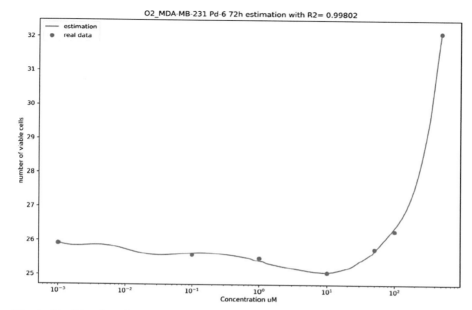

Fig. 4. The R2 value of produced superoxide anion radical of MDA-MB-231 cells treated with Pd-6 after 72 h.

This solution is a universal mathematical function that can be used in all platforms and programs that have these mathematical functions. The parsimony coefficient was adjusted based on the desired complexity and precision of the solution, which penalizes an increase in the size of the function without a significant improvement in quality. It was also crucial to use a logarithmic scale during training because the data were distributed in such a way that using a linear scale would have yielded poor results at higher concentrations. Overall, the results demonstrate that GP is a powerful tool for creating mathematical models that accurately describe the relationship between concentration and the number of viable cells.

Fig. 5. Syntax tree of produced GSH from MDA-MB-231 cells treated with Pd5-Pd6 after 72 h with R2 value of 0.99455.

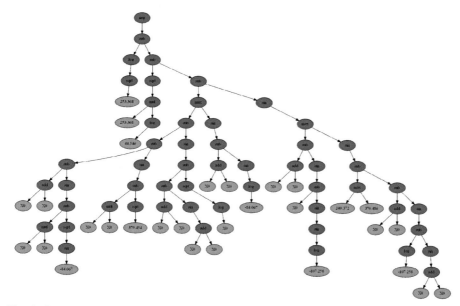

Fig. 6. Syntax tree of produced NO_2^- from MDA-MB-231 cells treated with Pd-5 after 72 h with an R2 value of 0.9761.

4   Conclusion

This study demonstrates that GP can be effectively applied to create mathematical models with high precision, as evidenced by R2 values ranging from 0.97 to 0.999. Utilizing such a method could streamline the process of determining appropriate concentrations for experimental use, potentially enhancing the efficiency of research in the field of cancer therapeutics. To further enhance this approach, it is suggested to expand the dataset and measure values at various time points, allowing GP to be trained with an additional time input. With only two time points of 24 h and 72 h available in the current study, a more comprehensive temporal function could not be generated. By incorporating a wider range of concentrations and time points, a more robust and versatile mathematical model could be developed. This enhanced model would provide a more comprehensive understanding of the relationship between concentration and the number of viable cells, ultimately contributing to the optimization of treatment strategies for various types of cancer.

Acknowledgments. This research is supported by the project that has received funding from the European Union's Horizon 2020 research and innovation programmes under grant agreement No 952603 (SGABU project). The research was funded by the Ministry of Education, Science and Technological Development of the Republic of Serbia, contract number [451-03-68/2022-14/200378 (Institute of Information Technologies, University of Kragujevac) and 451-03-68/2022-14/200107 (Faculty of Engineering, University of Kragujevac)].

References

Achkar, I.W., Abdulrahman, N., Al-Sulaiti, H., Joseph, J.M., Uddin, S., Mraiche, F.: Cisplatin based therapy: the role of the mitogen activated protein kinase signaling pathway. J. Transl. Med. **16**(1), 96 (2018). https://doi.org/10.1186/s12967-018-1471-1

Aldossary, S.A.: Review on pharmacology of cisplatin: clinical use, toxicity and mechanism of resistance of Cisplatin. Biomed. Pharmacol. J. **12**(1), 07–15 (2019). https://doi.org/10.13005/bpj/1608

Brown, A., Kumar, S., Tchounwou, P.B.: Cisplatin-based chemotherapy of human cancers. J. Cancer Sci. Therapy **11**(4), 97 (2019)

Dabhi, V.K., Chaudhary, S.: Empirical modeling using genetic programming: a survey of issues and approaches. Nat. Comput.Comput. **14**(2), 303–330 (2015). https://doi.org/10.1007/s11047-014-9416-y

De Jong, K.: Learning with genetic algorithms: an overview. Mach. Learn. **3**(2–3), 121–138 (1988). https://doi.org/10.1007/BF00113894

Di Bucchianico, A.: Coefficient of Determination (R^2). In: Ruggeri, F., Kenett, R.S., Faltin, F.W. (eds.) Encyclopedia of Statistics in Quality and Reliability, 1st edn. Wiley (2007). https://doi.org/10.1002/9780470061572.eqr173

Eggermont, J.: A Division of Image Processing, Department of Radiology C2S, Leiden University Medical Center, P.O. Box 9600, 2300 RC Leiden, The Netherlands (2007)

Feizi-Dehnayebi, M., Dehghanian, E., Mansouri-Torshizi, H.: A novel palladium(II) antitumor agent: Synthesis, characterization, DFT perspective, CT-DNA and BSA interaction studies via in-vitro and in-silico approaches. Spectrochim. Acta Part A Mol. Biomol. Spectrosc. **249**, 119215 (2021). https://doi.org/10.1016/j.saa.2020.119215

Filho, R.M., Lacerda, A., Pappa, G.L.: Explaining symbolic regression predictions. In: 2020 IEEE Congress on Evolutionary Computation (CEC), pp. 1–8 (2020). https://doi.org/10.1109/CEC48606.2020.9185683

Kamienny, P.-A., Lample, G., Charton, F.: End-to-end Symbolic Regression with Transformers (2022)

Kavakiotis, I., Tsave, O., Salifoglou, A., Maglaveras, N., Vlahavas, I., Chouvarda, I.: Machine learning and data mining methods in diabetes research. Comput. Struct. Biotechnol. J. **15**, 104–116 (2017). https://doi.org/10.1016/j.csbj.2016.12.005

Koza, J.R.: Genetic programming as a means for programming computers by natural selection. Stat. Comput. **4**(2) (1994). https://doi.org/10.1007/BF00175355

Liang, H., Sun, L., Wang, M., Yang, Y.: Deep learning with customized abstract syntax tree for bug localization. IEEE Access **7**, 116309–116320 (2019). https://doi.org/10.1109/ACCESS.2019.2936948

Mbugua, S.N., et al.: New Palladium(II) and Platinum(II) complexes based on pyrrole schiff bases: synthesis, characterization, X-ray structure, and anticancer activity. ACS Omega **5**(25), 14942–14954 (2020). https://doi.org/10.1021/acsomega.0c00360

Petrović, V.P., Živanović, M.N., Simijonović, D., Đorović, J., Petrović, Z.D., Marković, S.D.: Chelate N, O-palladium(ii) complexes: Synthesis, characterization and biological activity. RSC Adv. **5**(105), 86274–86281 (2015). https://doi.org/10.1039/C5RA10204A

Quade, M., Abel, M., Shafi, K., Niven, R.K., Noack, B.R.: Prediction of dynamical systems by symbolic regression. Phys. Rev. E **94**(1), 012214 (2016). https://doi.org/10.1103/PhysRevE.94.012214

França, F., Ribeiro, C.: 6th Brazilian Symposium on Neural Networks (SBRN 2000), Rio de Janiero, Brazil, 22–25 November 2000 (2000)

Sheth, S., Mukherjea, D., Rybak, L.P., Ramkumar, V.: Mechanisms of cisplatin-induced ototoxicity and otoprotection. Front. Cell. Neurosci. **11**, 338 (2017). https://doi.org/10.3389/fncel.2017.00338

Tchounwou, P.B., Dasari, S., Noubissi, F.K., Ray, P., Kumar, S.: Advances in our understanding of the molecular mechanisms of action of cisplatin in cancer therapy. J. Exp. Pharmacol. **13**, 303–328 (2021). https://doi.org/10.2147/JEP.S267383

Lee, W., Kim, H.-Y.: Genetic algorithm implementation in Python. In: Fourth Annual ACIS International Conference on Computer and Information Science (ICIS 2005), pp. 8–11 (2005). https://doi.org/10.1109/ICIS.2005.69

Zhang, Q., Barri, K., Jiao, P., Salehi, H., Alavi, A.H.: Genetic programming in civil engineering: advent, applications and future trends. Artif. Intell. Rev. **54**(3), 1863–1885 (2021). https://doi.org/10.1007/s10462-020-09894-7

Langdon, W.B., Poli, R.: Foundations of genetic programming. Springer Science & Business Media (2013)

Application of Artificial Intelligence for Predicting of New Potential Inhibitors of Vitamin K Epoxide Reductase

Marko R. Antonijević[1]([✉]), Dejan A. Milenković[1], Edina H. Avdović[1],
and Zoran S. Marković[1,2]

[1] Institute for Information Technologies, Kragujevac, University of Kragujevac, Jovana Cvijića bb, 34000 Kragujevac, Serbia
{mantonijevic,dejanm,zmarkovic}@uni.kg.ac.rs,
edina.avdovic@pmf.kg.ac.rs

[2] Department of Natural Science and Mathematics, State University of Novi Pazar, Vuka Karadžića bb, 36300 Novi Pazar, Serbia

Abstract. Artificial intelligence (AI) integration in drug development has altered the pharmaceutical industry by delivering innovative and effective solutions to long-standing difficulties in this field. AI techniques and tools, such as machine learning and deep learning algorithms, can analyze enormous datasets and identify complex patterns, therefore speeding up the drug discovery process. The research presented in this paper introduces a systematic approach to the design of novel anticoagulative agents with significant therapeutic potential, driven by the utilization of AI in drug design and development. Coumarins are compounds isolated from plants, and they are responsible for a wide range of biological functions. The anticoagulant potential of coumarin derivatives is one of their most prominent pharmacological properties. Warfarin (**WFR**), a common oral anticoagulant, belongs to the 4-hydroxycoumarin class. In one of our most recent studies, we performed pharmacological profiling and anticoagulant activity testing of a (*E*)-3-(1-((4-hydroxy-3-methoxyphenyl)amino)ethylidene)-2,4-dioxochroman-7-yl acetate [15]. When compared to the **WFR**, it was discovered that the examined molecule has better anticoagulative capability, as well as a more desirable pharmacokinetic profile than **WFR**. Furthermore, preliminary laboratory tests revealed that (*E*)-3-(1-((4-hydroxy-3-methoxyphenyl)amino)-ethylidene)chromane-2,4-dione (**L**) has similar pharmacokinetic properties with increased water solubility, making it an even better candidate for the development of potential anticoagulants. By utilization of the CReM web server, a series of 1000 derivatives of **L** were generated and narrowed down to 46 compounds through screening based on drug-likeness rules. Toxicity assessments further refined the selection to 16 non-toxic candidates, subsequently subjected to molecular docking simulations. Four compounds emerged from this subset, demonstrating excellent inhibition of VKOR, and VKORC1, compared to the conventional anticoagulant **WFR**. Additional investigation through molecular dynamics simulations provided insights into their activity in a specific timeframe confirming the results obtained from molecular docking simulations. The obtained results strongly suggest that investigated compounds have shown promising anticoagulative potential and should be further tested.

© The Author(s), under exclusive license to Springer Nature Switzerland AG 2024
N. Filipović (Ed.): AAI 2023, LNNS 999, pp. 170–184, 2024.
https://doi.org/10.1007/978-3-031-60840-7_21

1 Introduction

The wide array of compounds currently employed as pharmacological agents used in various fields of medicine and science can be traced back to their origins in the natural world. Due to the inability to change location "at will", plants needed to develop compound production systems that allowed them to resist environmental stressors [1–3]. People quickly realized that plants produce substances that protect them from the harmful effects of the outside world, and "traditional medicine" that uses plants and their extracts in the treatment of many diseases was born. A book on roots and herbs called "Pen Tsao" by Shen Nong, which was written in 2900 BC, is considered the oldest pharmacopoeia in the world. It contains a description of 365 herbal medicines and drugs, many of which are still in use today [1–4]. Treatments with plant materials have long been an irreplaceable and almost the only form of treatment for various diseases. Many drugs in modern medicine, such as morphine, which is obtained by cutting poppy seeds and is used as a powerful analgesic, have their origins in traditional medicine [5]. Thanks to the technological and industrial revolution, scientists in the last 100 years have been able to identify numerous molecules responsible for various biological or physiological functions in plant extracts, and then modify them or apply them in modern medicine and the pharmaceutical industry [6–9]. However, obtaining active compounds with adequate pharmacokinetic and toxicological properties was time- and resource-consuming with a high impact on the environment. During the period prior to the advent of artificial intelligence (AI), the field of drug design predominantly relied on empirical methodologies and high-throughput screening techniques. These methods generally led to protracted and resource-intensive procedures, yielding outcomes with a restricted rate of success, with scientists often having to rely on chemical intuition and experience when designing the experiment. Accurately predicting molecular interactions, which are crucial for drug-target binding, posed significant challenges. The lack of comprehensive databases and computational tools has impeded the effective investigation of the chemical universe [10–12].

1.1 Incorporation of Artificial Intelligence in Drug Design

The incorporation of artificial intelligence in drug development has transformed the pharmaceutical industry by providing novel answers to long-standing problems. AI approaches, such as machine learning and deep learning algorithms, are capable of analyzing large datasets and discovering complicated patterns, hence speeding up the drug discovery process. Among the applications where AI excels are the *in silico* screening of compound libraries, prediction of drug-target interactions, and virtual screening for prospective lead compounds. Furthermore, AI plays an important role in target identification and validation, allowing for the discovery of novel therapeutic targets and their relevance to specific diseases. AI aids in the optimization of clinical trial design and patient stratification for customized medication, increasing clinical development efficiency and success rates. Furthermore, AI aids in the repurposing of current medications for new purposes, the discovery of novel therapeutic uses, and the possible reduction of time and cost associated with bringing new treatments to market. Even though the AI drug design and development is not nearly as perfect and has some important shortcomings, it allows for faster and more precise work, moving the workload from the lab to the computer [12–14].

Some of our previous endeavors involved the development of coumarin derivatives for various purposes, but due to limitations of laboratory conditions, time and resources, large-scale screening was not possible. However, with the development of AI, several hundred compounds can be generated and tested without the need for expensive equipment and chemicals, being highly efficient and environmentally friendly [12–15].

1.2 Coumarins, Warfarin, Genome-Induced Resistance and Anticoagulative Activity

Coumarins are one of the most widespread classes of compounds extracted from plants and are accountable for an extensive array of biological functions. They comprise a class of both natural and synthetic compounds that are characterized by a structural motif that is founded upon the benzopyrone skeleton. Coumarin-based compounds are widely distributed in the natural world, with a particular emphasis on plants, where more than 1300 have been isolated and identified [16–18]. They are frequently detected in cassia cinnamon, fragrant woodruff, and Tonka bean, among other plant species. Although their significance in the aforementioned plant species is considerable, their function is occasionally overlooked. Certain coumarin derivatives play a role in managing respiration, photosynthesis, and growth. Additionally, certain coumarin derivatives can be classified as phytoalexins due to their production by plants as a means of defense against detrimental diseases. Furthermore, coumarins have the capacity to be produced as metabolites within the organisms of certain microorganisms, as well as higher animals, apart from plants [16–20]. The investigation into coumarin and its derivatives commenced over two hundred years ago, specifically in 1820, when A. Vogel successfully extracted 2H-chromen-2-one (Fig. 1) from the fruit of the *Coumarouna odorata Aube plant* (also known as *Dipetrik odorata*). This substance is in modern science recognized as coumarin and derives its name from the plant from which it was originally extracted (*coumarou-tree*; in the language of the South American Indians of French Guiana).

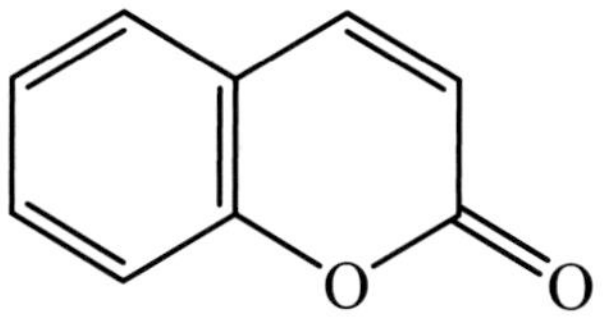

Fig. 1. Structure of the 2H-chromen-2-one, also known as coumarin.

Coumarin derivatives have a diverse set of physiological effects, which is why plant extracts containing them are frequently used to treat intestinal stains, typhoid, and leukemia. In addition, coumarins have been produced and modified for therapeutic applications, demonstrating a wide range of pharmacological properties. Coumarin derivatives often show high antioxidative potential [8, 15, 21–24], while some of their transition metal complexes represent promising anticancer agents [25–27]. They have antibacterial, antifungal, antiviral, antihyperglycemic, and anti-HIV properties [17–20, 28].

One of the most important pharmacological features of coumarin derivatives is their anticoagulative potential. Warfarin (**WFR**), a commonly used oral anticoagulant, belongs to the class of 4-hydroxycoumarins (Fig. 2). This compound, as well as other anticoagulative agents, expresses anticoagulative potential by inhibiting the enzyme vitamin K epoxide reductase (VKOR), resulting in a decrease in the active form of vitamin K. As a result, the synthesis of functional clotting factors II, VII, IX, and X is impaired, ultimately blocking the coagulation cascade.

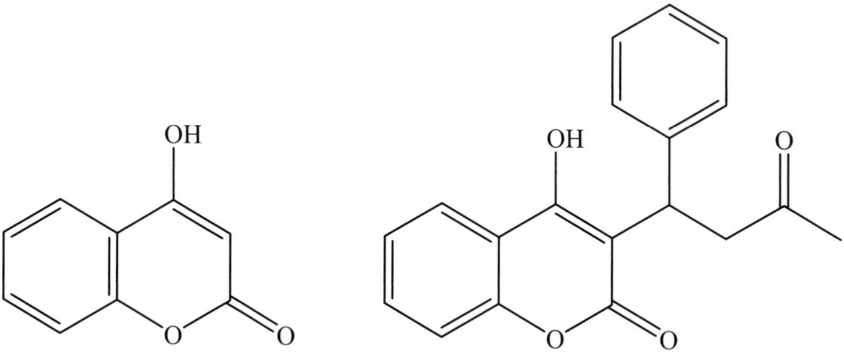

Fig. 2. Structure of the 4-hydroxycoumarin (left) and warfarin (right).

Coumarins' anticoagulant action has important therapeutic implications, particularly in the prevention and treatment of thromboembolic diseases. These compounds are commonly used for problems such as deep vein thrombosis, atrial fibrillation, and issues with prosthetic heart valves. Coumarins' capacity to control the coagulation process makes them effective instruments in the management of conditions where clot formation is a problem [29].

The main shortcomings which limit the usage of **WFR** in therapy, and reasons why scientists are looking for alternative solutions, are the relative toxicity of **WFR** and the occurrence of **WFR** resistance [30–32]. **WFR** resistance is characterized by a decrease in the anticoagulant medicine warfarin's ability to block the clotting cascade. This phenomenon is primarily caused by hereditary causes, most notably changes in the genes encoding the enzymes involved in the vitamin K cycle. Polymorphisms in the VKORC1 (Vitamin K epoxide reductase complex subunit 1) and CYP2C9 (Cytochrome P450 2C9) genes are well-known **WFR** resistance factors. VKORC1 encodes the target enzyme of warfarin, while CYP2C9 is in charge of the drug's metabolism. VKORC1 and CYP2C9 genetic variations can change the kinetics of warfarin metabolism and its interaction with the target enzyme, resulting in lower therapeutic efficacy. The c.-1639G > A polymorphism is the most common VKORC1 variant associated with **WFR** resistance, and it leads to increased production of the VKORC1 enzyme, requiring greater **WFR** doses to have the desired anticoagulant effect. Reduced enzyme activity caused by CYP2C9 variations, notably *2 and *3 alleles, affects warfarin metabolism and clearance.

Apart from genetic reasons, acquired **WFR** resistance mechanisms can also occur. These include increased vitamin K consumption, changes in hepatic metabolism, and medication interactions that impact the pharmacokinetics or dynamics of warfarin. Patients who are resistant to warfarin may require more frequent monitoring, greater doses, or alternate anticoagulant treatments to achieve therapeutic anticoagulation [30–32].

Understanding the hereditary and acquired variables that contribute to **WFR** resistance is critical for optimizing anticoagulant medication and reducing the chances of thromboembolic events or bleeding problems. Advances in pharmacogenomics have resulted in the development of dosage algorithms that use genetic information to estimate customized warfarin requirements, improving therapeutic precision and minimizing resistance issues.

In our previous endeavors, a series of coumarin derivatives with a focus on different biological activities were synthesized. As coumarin derivatives are well-known anticoagulative agents, some of them were subjected to comprehensive *in silico* and *in vitro* investigations, yielding some interesting results. For example, in one of our most recent papers, pharmacological profiling and investigation of the anticoagulant activity of the newly synthesized coumarin derivative: (*E*)-3-(1-((4-hydroxy-3-methoxyphenyl)amino)ethylidene)-2,4-dioxochroman-7-yl acetate was performed [15]. The obtained results were compared with the parameters obtained for **WFR**, and it was found that the investigated compound shows better anticoagulative potential, coupled with an adequate pharmacokinetic profile. Moreover, preliminary laboratory tests indicated that compound (*E*)-3-(1-((4–hydroxy-3-methoxyphenyl)amino)-ethylidene)chromane-2,4–dione (**L**), shows similar pharmacokinetic properties with increased water solubility which makes it even more suitable candidate for the development of potential anticoagulative agents. The structures of these compounds are given in Fig. 3.

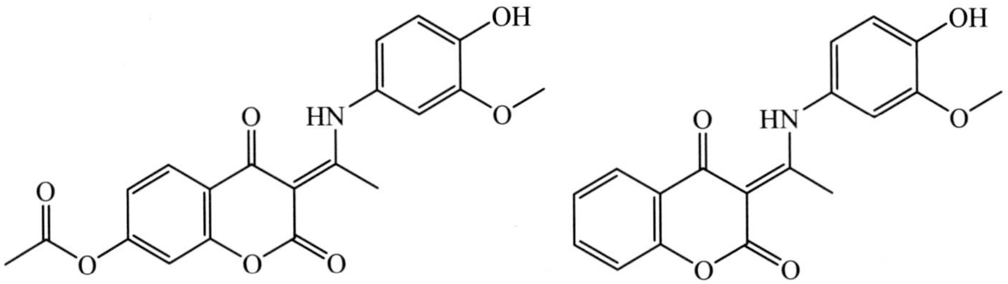

Fig. 3. Structures   of   (*E*)-3-(1-((4-hydroxy-3-methoxyphenyl)amino)ethylidene)-2,4-dioxochroman-7-yl   acetate   (left)   and   (*E*)-3-(1-((4–hydroxy-3-methoxyphenyl)amino)-ethylidene)chromane-2,4–dione (right).

In order to improve the efficacy, lower toxicity and bypass the resistance, while staying in accordance with principles of green and sustainable science, in this paper, artificial intelligence-based servers were employed to generate potential **WFR** substitute(s).

2 Methodology

For the design and development of novel anticoagulant agents as well as a better understanding of the anticoagulative potential of the investigated compounds, the deep learning models of binding affinity prediction were utilized. The Chemically Reasonable Mutations (CReM) webserver was employed to create a series of ligand (**L**) derivatives. CReM, an open-source Python framework that uses a fragment-based approach to generate chemical structures [33] was used to modify **L** and 1000 different synthetically acceptable **L** derivatives were obtained. To select viable candidates from the 1000 generated compounds, they are screened for anticoagulative potential, pharmacological properties, and toxicological profile.

Firstly, derivatives generated through CReM were subjected to ADMET properties evaluation. For this purpose, the ADMETlab 2.0 webserver was used. All derivatives of **L** which fulfill Pfizer Rule [34] and are found to be non-toxic were further chosen for further virtual screening. The selected non-toxic derivatives, **L** and warfarin (**WFR**) were subjected to molecular docking simulations in active binding sites of the Vitamin K Epoxide Reductase (VKOR). The three-dimensional crystal structure of VKOR is downloaded from the Protein Data Bank (PDB ID: 6WV3). UCSF Chimera was used to prepare the VKOR structure for virtual screening. Molecules of water and cofactor were removed, and the purified structure of VKOR was saved as a pdb format. The same software, UCSF Chimera, was used to prepare the native ligand of VKOR, **WFR**. Also, using Chimera, a binding pocket of VKOR was generated. For virtual screening, the MolAICal software was utilized. MolAICal software combines neural networks (artificial intelligence) and classical programming to design 3D ligands in the pocket of disease targets [35]. The native bound ligand of VKOR was extracted and a binding pocket analysis was performed. After that, molecular docking was performed with the investigated compounds. The AutoDock Vina molecular docking software is implemented in MolAICal [36]. Following the screening, compounds with binding energy values lower than those of **WFR** were chosen for further investigation of their pharmacokinetic properties. The pkCSM (predicting small-molecule pharmacokinetic properties using graph-based signatures) [37] was employed to examine the pharmacokinetic properties. Also, to bypass the **WFR** resistance, AI-assisted molecular docking simulations were repeated at the same conditions with a mutated version of VKOR (VKORC1 – PDB ID: 6WV5).

Compounds which fulfilled all the aforementioned criteria (four of them) were further subjected to molecular dynamics (MD) simulation, in order to investigate the behavior of protein-ligand complexes (VKOR with investigated compounds) in the specific time-frame, and validate the findings obtained through the molecular docking simulations. For this purpose, AMBER22 software package with AMBER Force Field was utilized [38]. This methodological approach was found to adequately describe these and similar biological processes [15, 24, 39, 40].

The structure of the investigated compounds, as well as the VKOR protein, were obtained from molecular docking simulations and topological parameters were generated using the AMBER force field. The Charmm-GUI server is used to generate topologies, input parameters, and coordinate files for the compounds under investigation [41]. The TIP3P solvation model is used to introduce solvation effects on the researched protein-ligand complexes. The solvated system was neutralized with K^+ or Cl^- (0.15 M KCl) ions using the Monte Carlo Ion Placing Method using 126 K^+ and 127 Cl^- ions. The energy of produced protein-ligand complexes was minimized using the steepest descent algorithm over 50 000 steps and the conjugate gradient algorithm with a tolerance of up to 1000 kJ mol^{-1} nm^{-1}. The equilibration phase took place under NVT ensemble conditions. The MD production process was carried out in an NPT assembly using the SHAKE algorithm at a time scale of 20 ns, with the implementation of a Monte-Carlo barostat (P = 2 ps) and Berendsen thermostat (300 K, T = 1 ps). Furthermore, from MD output trajectories, Root Mean Square Deviation (RMSD), Radius of Gyration (Rg), Root Mean Square Fluctuation (RMSF), and HB number, are generated for studying system attributes during and after molecular dynamics simulations, including general stability and structural fluctuations. These parameters are utilized to calculate the stability and structural changes of the protein-ligand complex across the estimated period. According to the RMSD measurements, the time it took for the protein to attain stable conformation and reach equilibrium was around 10 ns. Simulations were done in a 20 ns timescale to ensure the reliability of the results.

3   Results and Discussion

The CReM webserver was utilized to produce a set of 1000 unique possible derivatives with good synthetic accessibility, beginning from the compound **L**. The compounds were subsequently submitted to a comprehensive evaluation of their pharmacokinetic and toxicological properties using ADMETlab 2.0, a widely recognized screening tool. Based on Pfizer's criterion, a total of 46 compounds were identified as drug-like structures. Among these compounds, 16 were shown to be non-toxic, prompting their selection for subsequent molecular docking simulations. It is noteworthy to acknowledge that all chosen non-toxic derivatives also adhere to the Lipinski rule, along with other drug-likeness standards outlined in the ADMETlab 2.0.

3.1   Molecular Docking Simulations

In this study, the deep learning model of binding affinity prediction was used to virtually screen de novo drugs based on molecular docking results. This was accomplished using the MolAICal. The native bound ligand of VKOR was extracted, and binding pocket analysis was performed. After that, re-docking was performed with the screened compounds, and the results are presented in Table 1. The most stable conformation inside the active site of each compound, including **L** and **WFR** are presented in Fig. 4. Also, molecular docking simulations have been done with **L** as the starting compound from which all other investigated compounds were derived, as well as with the native bound ligand of VKOR enzymes crystallized form (**WFR**).

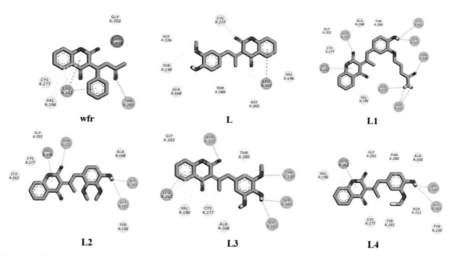

Fig. 4. Illustrative representation of the most stable conformations obtained through molecular docking simulations, and interactions between VKOR and investigated compounds. Green color represents hydrogen bonds, pink and violet π-alkil and π-σ interactions, yellow π-Sulphur interactions, and red unfavorable acceptor-acceptor interactions.

Table 1. The calculated values of free energies of binding of investigated compounds with VKOR enzyme

Ligands	ΔG_{bind} (kcal/mol)
WFR	−13.44
L	−13.59
L1	−14.19
L2	−13.57
L3	−13.09
L4	−13.07

As can be seen from Table 1, all investigated compounds, with the exception of **L1,** show similar binding affinity towards VKOR as **WFR** and **L**. Additionally, **L1** shows even better inhibitory potential than the other investigated compounds towards VKOR. This indicates the excellent binding potential of investigated compounds towards VKOR enzyme. Moreover, in order to conduct further analysis, four derivatives (**L1-L4**) were subjected to AI-assisted molecular docking simulations using the mutated form of VKORC1 (PDB ID: 6WV5) under identical conditions as previous simulations. These findings, which are displayed in Table 2, suggest that the binding affinity of **L** derivatives to VKORC1 is greater than that of **WFR**, which means that investigated compounds can be potentially used even with patients which express **WFR** resistance.

Table 2. The calculated values of free energies of binding of investigated compounds with VKORC1

Ligands	ΔG_{bind} (kcal/mol)
WFR	−10.86
L	−11.25
L1	−12.27
L2	−11.30
L3	−11.61
L4	−12.28

It is interesting to notice that compound **L1** shows increased binding affinity towards both, VKOR and VKORC1 enzyme, which makes it an especially important candidate for further investigations with the goal of designing novel and improved anticoagulative agents. To further investigate their anticoagulative properties and explain their behavior in a specific timeframe, shedding light on their mode of action, **L1-L4** are subjected to molecular dynamics simulations.

3.2   Molecular Dynamics Simulations

The results obtained through molecular docking simulations and ADMET analysis indicated that out of 1000 generated structural analogues, 4 have fulfilled the criteria to be a potential **WFR** replacement. With binding energies which are a little bit lower than **WFR**, some of these compounds have the potential to be active in lower concentrations with potentially lower toxic effects and regardless of the genotype induce **WFR** resistance. To investigate their inhibitory mechanism, we need to take a deeper look at the parameters generated from the trajectories of the production step in MD simulations.

As can be seen from the RMSD diagram given in Fig. 5, compounds **L1** and **L4** show no significant effect on the secondary structure of the protein showing only the slight stabilization of the protein's secondary structure before it reaches equilibrium. On the other hand, compounds **L2** and **L3** show a significant increase in RMSD values, indicating that inhibition of the VKOR by these compounds has the basis for the disruption of the secondary structure of the investigated enzyme. This increase in RMSD values indicates instability of VKOR-**L2** and VKOR-**L3** protein-ligand complexes.

In addition, a closer look at the RMSF values presented in Fig. 6 also indicates that the binding of **L3** increases the flexibility of the amino acid residues, indicating the loss of the hydrogen bonds and other non-covalent interactions which hold the structure of the VKOR. The highest effect can be seen in the regions with amino acids numbered from 120 to 140 and from 310 to 320, with a maximum fluctuation difference of 1.5 nm. Other investigated compounds do not induce significant changes in the fluctuation of amino acid residues.

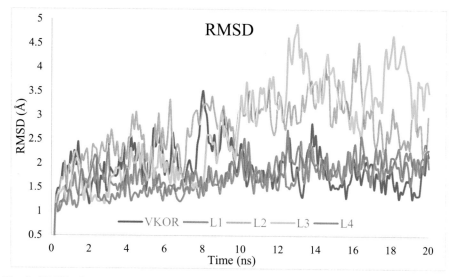

Fig. 5. RMSD diagram describing the changes in protein-ligand complex after binding **L1-L4** to VKOR.

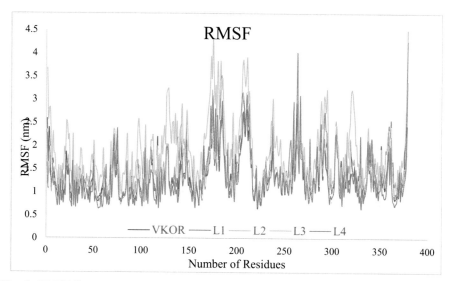

Fig. 6. RMSF diagram describing the changes in protein-ligand complex after binding **L1-L4** to VKOR.

Also, a significant difference in the values of radius gyration (Fig. 7) in comparison is also detected when the VKOR-**L3** complex is investigated. Namely, according to the results presented in Rg diagrams, binding of the **L3** induces an increase in the stability and compactness of the investigated protein, while other investigated compounds show little to no stabilizing effect. However, after reaching equilibrium, compounds **L1** and

L3 introduce the disruption in the structure of VKOR inducing reduced stability and compactness of the protein-ligand complex.

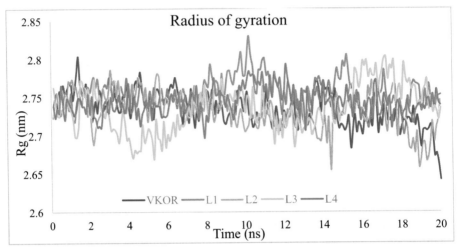

Fig. 7. Rg diagram describing the changes in protein-ligand complex after binding **L1-L4** to VKOR.

These trends set by binding **L1-L4** to VKOR and presented at the RMSD, RMSF, and Rg diagrams, as well as the results of molecular docking simulation can, in part, be explained through the number of hydrogen bonds maintained throughout the production step of MD simulations.

The average number of maintained hydrogen bonds for compounds **L1** and **L3** is higher than one, 1.68 and 1.02 respectively, indicating that throughout the simulation these compounds maintained strongly bonded to the VKOR. This is especially obvious when Fig. 8 is examined, where we can see that these compounds often have 3–4 hydrogen bonds simultaneously. On the other hand, compound **L2** partially follows the trend set by the previous compounds with an average number of HB of 0.80, while **L4** shows a low number of HB of 0.21, indicating that binding energies are a consequence of other non-covalent interactions, thus having a lower effect to the changes in the protein structure. This trend is in direct correlation with the number of HB obtained as a result of molecular docking simulations and presented in Fig. 4. In comparison to the **WFR** binding investigated in our previous work [15], **WFR** shows binding affinity similar to L, **L2**, and **L4** while **L1** and **L3** show stronger inhibitory potential according to the RMSD, RMSF and number of HB maintained throughout time. This means that **L1** and **L3** will have better anticoagulative potential, while the activity of **L2** and **L4** will be similar to that of **WFR**, which is in accordance with results obtained through molecular docking simulations. Because of that, all four AI-generated compounds should be synthesized and then subjected to *in vitro*, as well as *in vivo* biological and pharmacological tests. It is important to note that if compounds express too aggressive anticoagulative potential, they can induce problems such are internal or external bleeding, which can be the case for some of the investigated compounds. However, in that case, compounds with these

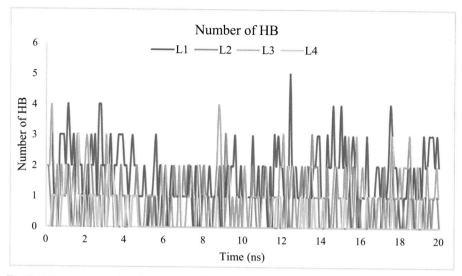

Fig. 8. Diagram presenting the number of hydrogen bonds in protein-ligand complex after binding **L1-L4** to VKOR.

properties will be characterized as "Superwarfarins" and could potentially be utilized in alternative industrial applications, such as rodenticides [42].

Despite their potential therapeutic efficacy, coumarins must be used with caution due to their narrow therapeutic window and the possibility of interactions with other medicines and dietary variables. Maintaining a balance between anticoagulation and the risk of bleeding is critical for maximizing coumarin therapeutic advantages, and potential anticoagulative agents should be thoroughly *in vitro* and *in vivo* tested before further utilization.

4   Conclusion

In conclusion, in this study we have employed a systematic methodology based on artificial intelligence to design novel anticoagulative agents with promising potential. Leveraging the CReM web server, 1000 derivatives of **L** were generated, and subsequent screening based on drug-likeness rules narrowed down the selection to 46 compounds. Rigorous toxicity assessments identified 16 non-toxic candidates, which were further evaluated through molecular docking simulations. From this subset, four compounds emerged as more potent inhibitors of VKOR than the widely used warfarin. These promising candidates underwent additional investigation through molecular dynamics simulations, revealing their mode of action and subsequently confirming the results obtained through molecular docking simulations. The robust computational analyses strongly suggest that these four novel anticoagulative agents hold substantial potential. Considering the *in silico* obtained results, it is endorsed that these compounds be subjected to *in vitro* testing to validate their efficacy and safety profiles. This marks a crucial step in the translational journey of these agents from computational design to practical application in anticoagulation therapy and demonstrates the utilization of AI in drug design and development. The results obtained here pave the way for further exploration and development of novel anticoagulative agents as well as other medications using AI.

Acknowledgment. The authors wish to acknowledge the Ministry of Education, Science, and Technological Development of the Republic of Serbia (Agreements No. 451-03-47/2023-01/200378 and 451-03-47/2023-01/200252) for financial support.

References

1. Rafieian-Kopaei, M., Baradaran, A., Rafieian, M.: Plants antioxidants: From laboratory to clinic. J. Nephropathology **2**(2), 152 (2013)
2. Andre, C.M., Larondelle, Y., Evers, D.: Dietary antioxidants and oxidative stress from a human and plant perspective: a review. Curr. Nutr. Food Sci. **6**(1), 2–12 (2010)
3. Pérez-Torres, I., Castrejón-Téllez, V., Soto, M.E., Rubio-Ruiz, M.E., Manzano-Pech, L., Guarner-Lans, V.: Oxidative stress, plant natural antioxidants, and obesity. Int. J. Mol. Sci. **22**(4), 1786 (2021)
4. Du, G.H., Yuan, T.Y., Zhang, Y.X.: The potential of traditional Chinese medicine in the treatment and modulation of pain. Adv. Pharmacol. **75**, 325–361 (2016)
5. Hedayati-Moghadam, M., et al.: The effects of Papaver somniferum (Opium poppy) on health, its controversies and consensus evidence. Toxin Rev. **41**(3), 1030–1043 (2022)
6. Dimić, D., Milenković, D., Marković, J.D., Marković, Z.: Antiradical activity of cate-cholamines and metabolites of dopamine: theoretical and experimental study. Phys. Chem. **19**(20), 12970–12980 (2017)
7. Milenković, D., Đorović, J., Jeremić, S., Dimitrić Marković, J.M., Avdović, E.H., Marković, Z.: Free radical scavenging potency of dihydroxybenzoic acids. J. Chem. (2017)
8. Milanović, ŽB., et al.: Inhibitory activity of quercetin, its metabolite, and standard antiviral drugs towards enzymes essential for SARS-CoV-2: The role of acid–base equilibria. RSC Adv. **11**(5), 2838–2847 (2021)
9. Yuan, H., Ma, Q., Ye, L., Piao, G.: The traditional medicine and modern medicine from natural products. Molecules **21**(5), 559 (2016)
10. Baldi, A.: Computational approaches for drug design and discovery: an overview. Systematic Rev. Pharmacy **1**(1), 99 (2010)
11. Honarparvar, B., Govender, T., Maguire, G.E., Soliman, M.E., Kruger, H.G.: Integrated approach to structure-based enzymatic drug design: molecular modeling, spectroscopy, and experimental bioactivity. Chem. Rev. **114**(1), 493–537 (2014)
12. Brown, N., Ertl, P., Lewis, R., Luksch, T., Reker, D., Schneider, N.: Artificial intelligence in chemistry and drug design. J. Comput. Aided Mol. Des. **34**, 709–715 (2020)
13. Haider, R.: Drug Discovery in the 21st Century. Toxi App. Pharma Insights Res. **6**(1), 65–76 (2023)
14. Jones, J. We Improve Health Through Innovative Solutions
15. Milanović, Ž., et al.: In silico evaluation of pharmacokinetic parameters, delivery, distribution and anticoagulative effects of new 4, 7-dihydroxycoumarin derivative. J. Biomolecul. Struct. Dyn., 1–16 (2023)
16. Murray, R.D.H.: Coumarins. Natural Product Rep. **6**(6), 591–624 (1989)
17. Vazquez-Rodriguez, S., Joao Matos, M., Borges, F., Uriarte, E., Santana, L.: Bioactive coumarins from marine sources: origin, structural features and pharmacological properties. Curr. Top. Med. Chem. **15**(17), 1755–1766 (2015)

18. Riveiro, M.E., et al.: Coumarins: old compounds with novel promising therapeutic perspectives. Curr. Med. Chem. **17**(13), 1325–1338 (2010)
19. Srikrishna, D., Godugu, C., Dubey, P.K.: A review on pharmacological properties of coumarins. Mini Rev. Med. Chem. **18**(2), 113–141 (2018)
20. Sharifi-Rad, J., et al.: Natural coumarins: exploring the pharmacological complexity and underlying molecular mechanisms. Oxidative Medicine and Cellular Longevity (2021)
21. Antonijević, M.R., et al.: Green one-pot synthesis of coumarin-hydroxybenzohydrazide hybrids and their antioxidant potency. Antioxidants **10**(7), 1106 (2021)
22. Antonijević, M.R., Avdović, E.H., Simijonović, D.M., Milanović, ŽB., Amić, A.D., Marković, Z.S.: Radical scavenging activity and pharmacokinetic properties of coumarin–hydroxybenzohydrazide hybrids. Int. J. Mol. Sci. **23**(1), 490 (2022)
23. Simijonović, D.M., et al.: Coumarin N-acylhydrazone derivatives: green synthesis and antioxidant potential—experimental and theoretical study. Antioxidants **12**(10), 1858 (2023)
24. Vasić, J., et al.: The electronic effects of 3-methoxycarbonylcoumarin substituents on spectral, antioxidant, and protein binding properties. Int. J. Mol. Sci. **24**(14), 11820 (2023)
25. Dimić, D.S., et al.: Synthesis, Crystallographic, quantum chemical, antitumor, and molecular docking/dynamic studies of 4-hydroxycoumarin-neurotransmitter derivatives. Int. J. Mol. Sci. **23**(2), 1001 (2022)
26. Avdović, E.H., et al.: Synthesis and biological screening of new 4-hydroxycoumarin derivatives and their palladium (II) complexes. Oxid. Med. Cell. Longev. **2021**, 1–18 (2021)
27. Avdović, E.H., et al.: Synthesis and cytotoxicity evaluation of novel coumarin-palladium (II) complexes against human cancer cell lines. Pharmaceuticals **16**(1), 49 (2022)
28. Milenković, D.A., Dimić, D.S., Avdović, E.H., Marković, Z.S.: Several coumarin derivatives and their Pd (II) complexes as potential inhibitors of the main protease of SARS-CoV-2, an in silico approach. RSC Adv. **10**(58), 35099–35108 (2020)
29. Kumar, A., Kumar, P., Shravya, H., Pai, A.: Coumarins as potential anticoagulant agents. Res. J. Pharmacy Technol. **15**(4), 1659–1663 (2022)
30. Oldenburg, J., Müller, C.R., Rost, S., Watzka, M., Bevans, C.G.: Comparative genetics of warfarin resistance. Hamostaseologie **34**(02), 143–159 (2014)
31. Rost, S., et al.: Mutations in VKORC1 cause warfarin resistance and multiple coagulation factor deficiency type 2. Nature **427**(6974), 537–541 (2004)
32. Hulse, M.L.: Warfarin resistance: diagnosis and therapeutic alternatives. Pharmacotherapy: J. Hum. Pharmacol. Drug Therapy **16**(6), 1009–1017 (1996)
33. Polishchuk, P.: CReM: chemically reasonable mutations framework for structure generation. J. Cheminform. **12**(1), 1–18 (2020)
34. Hughes, J.D., et al.: Physiochemical drug properties associated with in vivo toxicological outcomes. Bioorg. Med. Chem. Lett. **18**(17), 4872–4875 (2008)
35. Bai, Q., Tan, S., Xu, T., Liu, H., Huang, J., Yao, X.: MolAICal: a soft tool for 3D drug design of protein targets by artificial intelligence and classical algorithm. Briefings Bioinform. **22**(3), bbaa161 (2021)
36. Trott, O., Olson, A.J.: AutoDock Vina: improving the speed and accuracy of docking with a new scoring function, efficient optimization, and multithreading. J. Comput. Chem. **31**(2), 455–461 (2010)
37. Pires, D.E., Blundell, T.L., Ascher, D.B.: PkCSM: predicting small-molecule pharmacokinetic and toxicity properties using graph-based signatures. J. Med. Chem. **58**(9), 4066–4072 (2015)
38. Case, D.A., et al.: Amber 2021. University of California, San Francisco (2021)
39. Jovanović, J.Ð, Antonijević, M., Vojinović, R., Filipović, N.D., Marković, Z.: In silico study of inhibitory capacity of sacubitril/valsartan toward neprilysin and angiotensin receptor. RSC Adv. **12**(46), 29719–29726 (2022)

40. Jovanović, J.Đ, Antonijević, M., El-Emam, A.A., Marković, Z.: Comparative MD Study of Inhibitory Activity of Opaganib and Adamantane-Isothiourea Derivatives toward COVID-19 Main Protease Mpro. ChemistrySelect **6**(33), 8603–8610 (2021)
41. Jo, S., Kim, T., Iyer, V.G., Im, W.: CHARMM-GUI: a web based graphical user interface for CHARMM. J. Comput. Chem. **29**(11), 1859–1865 (2008)
42. Murphy, M.J., Lugo, A.M.: Superwarfarins. In: Handbook of Toxicology of Chemical Warfare Agents, pp. 207–223. Academic Press (2009)

Predicting the Evolution of Cancer Stem Cell Subtypes Using a Machine Learning Framework

D. Nikolić[1,2](✉), B. Ljujić[3], A. Ramović Hamzagić[3], M. Gazdić Janković[3], A. Mirić[1],
K. Virijević[1], D. Šeklić[1], M. Jovanović[4], N. Kastratović[3], I. Petrović[3], V. Jurišić[3],
N. Milivojević[1], M. Živanović[1], and N. Filipović[2,5]

[1] Institute for Information Technologies, University of Kragujevac, Jovana Cvijića bb,
34000 Kragujevac, Serbia
markovac85@kg.ac.rs

[2] BioIRC - Bioengineering Research and Development Center, University of Kragujevac,
Prvoslava Stojanovića 6, 34000 Kragujevac, Serbia
fica@kg.ac.rs

[3] Faculty of Medical Sciences, University of Kragujevac, Svetozara Markovića 69,
34000 Kragujevac, Serbia

[4] Faculty of Sciences, University of Kragujevac, Radoja Domanovića 12, 34000 Kragujevac,
Serbia
milena.jovanovic@pmf.kg.ac.rs

[5] Faculty of Engineering, University of Kragujevac, Sestre Janjić 6, 34000 Kragujevac, Serbia

Abstract. Cancer stem cells (CSCs) are subpopulation of cells in a tumor that are very important for analysis and treatment in clinical practice. The aim of this study is the use of Machine learning (ML) methodology to predict the development of CSCs subpopulation in colon and breast cancer cells. Input data for training Genetic algorithm (GA) and fitting was used from experimental measurements on flow cytometry of CSCs surface markers expression in cancer cells. Based on the results, GA prediction model has archived high accuracy in estimating the expression rate of CSCs markers on cancer cells. Artificial intelligence can be used as a powerful tool for predicting of behavior of cancer stem cell subpopulation.

Keywords: Machine learning · Genetic Algorithm · Cancer Stem Cells · Flow cytometry

1 Introduction

Cancer Stem Cells (CSCs) stand at the forefront of tumor biology, representing a pivotal subpopulation of cells with unparalleled influence over the initiation, progression, metastasis, and recurrence of malignancies [1]. The significance of understanding CSCs lies in their unique ability to govern the diverse facets of tumor behavior, making them compelling targets for therapeutic intervention and furthering our comprehension of cancer as a dynamic and heterogeneous disease.

The identification of CSCs poses a formidable challenge due to their elusive nature and the intricate interplay of various surface markers that define their identity. Recent

N. Filipović (Ed.): AAI 2023, LNNS 999, pp. 185–189, 2024.
https://doi.org/10.1007/978-3-031-60840-7_22

advancements have illuminated a plethora of such markers, underscoring the need for a nuanced approach that involves the simultaneous assessment of multiple markers to accurately pinpoint CSCs [2–4]. This multifaceted challenge necessitates innovative methodologies that transcend traditional boundaries, and in this pursuit, mathematical models, particularly those employing Machine Learning (ML) techniques, emerge as indispensable allies.

Machine Learning, with its capacity to discern patterns and relationships within complex datasets, takes center stage in the endeavor to unravel the intricacies surrounding CSCs. In this context, the Genetic Algorithm (GA) serves as a potent ML approach, drawing inspiration from the principles of natural evolution articulated by Charles Darwin [5]. GA, functioning as a metaheuristic method, provides a unique advantage in the exploration of the vast solution space inherent in the identification and characterization of CSCs.

Our study delves into this synergistic intersection of CSC biology and ML methodologies, with a specific focus on the application of Genetic Algorithm. The inherent complexity of CSC biology demands not only an understanding of individual markers but also an exploration of the dynamic relationships and hierarchies within CSC subpopulations. GA, through its iterative refinement process mirroring natural selection, becomes a powerful instrument for navigating this complexity.

Genetic Algorithm's ability to generate high-quality solutions for optimization and search problems positions it as an innovative tool in the quest to unravel the mysteries surrounding CSCs. By mimicking the evolutionary process, GA adapts and refines solutions, providing insights into the intricate patterns within the expression profiles of multiple CSC markers. This not only enhances our ability to identify and characterize CSCs but also holds the promise of predicting the developmental trajectory of CSC subpopulations.

As we embark on this scientific journey, the fusion of biological insights with computational prowess takes on profound significance. The integration of ML methodologies, particularly GA, augments our analytical capabilities, enabling a comprehensive understanding of the nuanced landscape of CSCs. This not only informs our fundamental understanding of cancer biology but also holds transformative potential for clinical applications, ranging from improved diagnostics to the development of targeted therapeutic interventions.

In conclusion, the synergy between CSC biology and Machine Learning, exemplified by the utilization of Genetic Algorithm, represents a paradigm shift in cancer research. It offers a holistic approach to unraveling the complexities inherent in CSC identification and characterization. The knowledge derived from this integration not only enhances our grasp of cancer biology but also charts a course towards personalized and more effective cancer therapies, marking a significant stride towards a future where CSCs are not merely elusive adversaries but well-understood entities with targeted vulnerabilities.

2 Material and Methods

2.1 Flow Cytometry Analysis

The investigation of Cancer Stem Cells (CSCs) necessitates a meticulous exploration of their surface marker expressions, a task achieved through the utilization of flow cytometry. The analysis was executed using a BD Biosciences FACSCalibur, a sophisticated instrument renowned for its accuracy in cellular analysis. The acquired data were subsequently processed using the commercial software Flowing, known for its robust analytical capabilities.

Two distinct cancer cell lines, HCT-116 colorectal cancer and MDA-MB-231 breast cancer, both sourced from the European Collection of Authenticated Cell Cultures (ECACC), served as the primary subjects for this analysis. Cultivated in 6-well dishes, these cell lines provided a conducive environment for the screening of various cancer stem cell surface markers.

To discern the expression patterns of these crucial markers, the cells underwent incubation with anti-human monoclonal antibodies specifically designed to target and identify CSCs. This approach allowed for a comprehensive examination of the intricate surface marker landscape associated with cancer stem cells.

2.2 Machine Learning Model (ML)- Genetic Algorithm (GA)

In tandem with the experimental endeavors, a robust Machine Learning model, specifically the Genetic Algorithm (GA), was employed to unravel intricate patterns within the data. The input data for training the GA and subsequent fitting were derived from the experimental measurements obtained through flow cytometry, capturing the dynamic expressions of CSC markers in cancer cells.

The fundamental principles governing the ML-GA approach are elucidated and articulated by O'Neill et al. [6]. The integration of GA into the analytical framework introduces a layer of sophistication, as it mimics the principles of natural evolution to iteratively optimize and refine solutions. This iterative process aligns with the dynamic nature of CSCs, allowing the model to adapt and enhance its predictive capacity based on the nuances identified in the flow cytometry data.

The GA operates through the generation of diverse solutions, mirroring genetic variation, and subjecting them to selective pressures to emulate natural selection. This iterative refinement process is finely tuned to converge on optimal solutions, enhancing the model's ability to predict the developmental trajectory of CSC subpopulations.

In summary, the materials and methods employed in this study represent a seamless integration of cutting-edge experimental techniques, such as flow cytometry, with advanced computational methodologies, exemplified by the ML-GA model. This synergistic approach not only enables a comprehensive exploration of CSC surface markers but also empowers the model to discern intricate patterns within the data, propelling our understanding of CSC dynamics to new heights. The convergence of these methodologies positions our study at the forefront of cancer research, offering novel insights and paving the way for future advancements in precision medicine and targeted therapeutic interventions.

3   Results and Conclusions

3.1   Surface Marker Expression Analysis

The focal point of this study was the comprehensive analysis of Cancer Stem Cells (CSCs) surface markers, including CD24, CD44, ALDH1, and ABCG2, within two distinct cancer cell lines - HCT-116 colorectal cancer and MDA-MB-231 breast cancer. Through meticulous flow cytometry analysis, we elucidated the dynamic expression patterns of these markers, unraveling the intricate landscape associated with CSCs in these particular cancer contexts.

Our findings revealed nuanced variations in the expression profiles of CD24, CD44, ALDH1, and ABCG2 across the two cancer cell lines. Notably, CD44 exhibited heightened expression in the HCT-116 colorectal cancer cells, suggesting a potential subtype-specific association with colorectal cancers. Conversely, CD24 showed a more pronounced expression in the MDA-MB-231 breast cancer cells. These subtle variations underscore the heterogeneity within CSC subpopulations, emphasizing the need for a comprehensive approach to understanding and targeting these cells.

3.2   Machine Learning - Genetic Algorithm Analysis

Parallel to our experimental efforts, the integration of the Genetic Algorithm (GA) and Machine Learning (ML) model presented a pioneering avenue for understanding the complexities inherent in CSC behavior. The GA, trained on experimental flow cytometry data, demonstrated a remarkable capacity to discern subtle patterns and relationships within the multidimensional landscape of CSC surface markers.

The GA, inspired by the principles of natural evolution, iteratively refined its predictive models to converge on optimal solutions. This adaptive process enabled the model to capture the temporal dynamics of CSC subpopulations, offering insights into the developmental trajectory of these cells over time. Notably, the GA identified correlations among specific surface markers, shedding light on potential regulatory networks governing CSC behavior.

3.3   Unique Contribution of *in Silico* Approaches

Our results emphasize the pioneering nature of the combined *in silico* and experimental methodology employed in this study. The integration of GA and ML techniques for the analysis of cancer stem cell markers represents an unexplored frontier in cancer research. The precision afforded by the GA in *in silico* testing, analysis, and monitoring of CSC behavior over time is a significant breakthrough.

This novel approach not only enhances our ability to interpret experimental results but also provides a predictive framework for understanding the intricate dynamics of CSC subpopulations. The GA's capacity to uncover hidden patterns within the data adds a layer of sophistication to our understanding of CSC biology, transcending traditional analytical boundaries.

3.4 Future Implications and Model System

The unique combination of in silico and experimental methodologies showcased in this study establishes a solid foundation for future research endeavors. The success of the GA in deciphering complex relationships within CSC surface markers positions it as a valuable tool for similar studies across different cancer types. The model system developed here serves as a blueprint for the integration of computational methods in the exploration of cancer stem cell biology.

As we navigate the uncharted waters of cancer research, the synergy between experimental and computational approaches becomes increasingly crucial. The predictive capabilities of the GA pave the way for tailored interventions, moving us closer to personalized medicine in the realm of cancer treatment. Additionally, the identified correlations among surface markers open avenues for targeted therapeutic strategies, marking a paradigm shift in the way we approach cancer therapy.

3.5 Conclusion

In conclusion, this study represents a significant leap forward in our understanding of Cancer Stem Cells and their surface markers. The integration of flow cytometry with the innovative ML-GA model provides a holistic view of CSC behavior, shedding light on the dynamic nature of these cells. The nuanced variations in surface marker expression across different cancer types underscore the heterogeneity within CSC subpopulations.

The predictive prowess of the GA not only refines our comprehension of CSC dynamics but also lays the groundwork for future advancements in precision medicine. As we move forward, the unique combination of in silico and experimental methodologies presented here sets a precedent for cutting-edge research in cancer biology. Our findings not only contribute to the current body of knowledge but also inspire new avenues for research, with the potential to revolutionize the way we diagnose, monitor, and treat cancer.

Acknowledgements. The research was funded by the Ministry of Science, Technological Development and Innovation of the Republic of Serbia, contract number [451-03-47/2023-01/200378 (Institute for Information Technologies Kragujevac, University of Kragujevac)].

References

1. Li, W., Ma, H., Zhang, J., et al.: Unraveling the roles of CD44/CD24 and ALDH1 as cancer stem cell markers in tumorigenesis and metastasis. Sci. Rep., vols. 7, 13856
2. Enciso-Benavides, J., et al.: Biological characteristics of a sub-population of cancer stem cells from two triple-negative breast tumour cell lines (2021). Heliyon, vol. 10; 7(6), e07273
3. Ginestier, C., et al.: ALDH1 is a marker of normal and malignant human mammary stem cells and a predictor of poor clinical outcome. Cell Stem Cell 1, 555–567 (2007)
4. Tiezzi, D.G., et al.: ABCG2 as a potential cancer stem cell marker in breast cancer. J. Clin. Oncol. **31**(15), e12007 (2013)
5. De Jong, K.: Learning with genetic algorithms: an overview. Mach. Learn. **3**(2), 121–138 (1988)
6. O'Neill, M., Poli, R., Langdon, W.B., McPhee Mach, N.F.: A Field Guide to Genetic Programming, p. 10, 229–230 (2009)

Digital Empathy

Dejan Maslikov ić[1](✉) and Đurađ Grubišić[2]

[1] Institut društvenih nauka, Kraljice Natlije 45, Beograd, Serbia
maslikovicd@idn.org.rs
[2] Evropski centar za mir i razvoj, Beograd, Serbia

Abstract. Robots with artificial intelligence have their place in medicine and patient care as therapists or patient assistants. This is particularly applicable to people with various types of disabilities, and the greatest application is for patients with various neurodegenerative pathologies (Parkinson's disease, Alzheimer's disease, multiple sclerosis, etc.). Also, artificial intelligence is gaining ground in gerontology and telemedicine.

The paper is based on the fact that the world's population, especially in developed countries, is getting older and that in the next few decades, almost a third of the population will be over 60, which makes it a challenge to provide care and medical services for such a large population. This led to the coining of the term social and emotional robotics. This segment of robotics should be in charge of creating a therapeutic assistant or even some kind of patient supervision.

So far, the greatest need has been expressed among people with disabilities—blind, visually impaired, wheelchair users, or those with neurodegenerative changes.

This direction of development and use of artificial intelligence opens up the question of empathy, emotional interaction, and social intelligence, i.e., whether it is possible to develop these features of artificial intelligence, which until now were only characteristic of living beings.

The paper will define certain problems with the use of artificial intelligence and interaction with people, that is, with users of services, in this case, patients. The importance of giving robots, i.e., artificial intelligence, the ability to understand emotions and interpret human behavior will be highlighted. In recent decades, in parallel with the development of robotics and artificial intelligence, neuroscience and cognitive psychology have been developed with the aim of investigating human emotions and behavior in detail. Important for science and AAI is the definition of artificial intelligence mechanisms that would allow robots to recognize emotions, facial expressions, gestures, and, finally, the mood of patients.

Keywords: medicine · gerontology · telemedicine · empathy · ethics · artificial intelligence · inclusion

This paper was written as part of the 2023 Research Program of the Institute of Social Sciences with the support of the Ministry of Science, Technological Development and Innovation of the Republic of Serbia.

N. Filipović (Ed.): AAI 2023, LNNS 999, pp. 190–193, 2024.
https://doi.org/10.1007/978-3-031-60840-7_23

1 Introduction

Artificial intelligence is higher, we can freely say that it is the highest level of intelligence that does not belong to the human race or any form of biological life on earth. We can distinguish several degrees of artificial intelligence by the degree of intelligence, which ranges from a tool that routinely performs strictly defined tasks to smart machines that, similar to human intelligence, develop the ability to learn.

Starting from the fact that artificial intelligence is defined as a way of reasoning and acting on executable conclusions completely based on logic, programmed and not human, and that the process of logically deriving conclusions is completely performed by a machine, it is necessary, following the views of Prof. Devillers, to carefully approach the development of the use of AI as an assistant for persons with disabilities or medical therapists.

The discussion in this paper will open up a new field for research that emerges from the integration of intention, empathy, and human-like creativity. It is extremely difficult for robots with artificial intelligence to make decisions on their own because they have no emotions, empathy, instinct, or intention to initiate decision-making.

Viewed from the point of view of science and the development of artificial intelligence, the use of robots in patient care and as assistants to people with disabilities has invaluable importance for the development of artificial intelligence and machine learning. The interaction between the human (patient) and the robot (artificial intelligence) is invaluable for science. This opens up a new dimension of the role of AI: learning to live in the community.

This also opens up space for the development and standardization of robots that must take on a human-like form. Movements, facial expressions, and gesticulation are not only indicators of emotions but also motor functions for the execution of actions.

In this paper, we try to identify important issues of the use of artificial intelligence: empathy, social integration, and interaction of machines and devices based on artificial intelligence with people, service users, and patients, as well as the responsibilities of artificial intelligence for decisions made, emotions expressed or attitudes expressed, the ethics of their actions, as well as the responsibility of the companies or individuals who programmed them.

The paper will not only address the important issue of legal regulations for the use of artificial intelligence but also the social effects of the application of devices and machines that use artificial intelligence. The absence of empathy, social interaction, and authority in areas of human activity and work such as education, medicine, social care, etc. are inadmissible.

2 Methodology

In the paper, a combination of theoretical-systemic and factographic methods will be used to collect facts, which will be used to describe and explain the phenomena and processes of importance for this topic. The mentioned methods and techniques will be used to investigate social phenomena (occurrences and processes) in different areas of the use of AI and their mutual relations, while the consequences will be considered in the discussion.

Also, the comparative method will be used as the most general method for comparing phenomena that occur in society using AI. Content analysis of published research results and published papers will contribute to the discussion on this important topic.

The case study will present several current projects involving the use of AI in medicine, telemedicine, and with persons with disabilities in Serbia.

3 Discussion

Currently, apart from the regulation of the use of personal data, society does not have a normative framework for the operation and use of machines and devices that use artificial intelligence. Also, at least publicly, no opinion has been expressed about the transition period and the consequences for society and the human species that will arise from the replacement of employees by devices and machines that use artificial intelligence.

The discussion will further deal with the legal regulation of the use of AI, responsibility, and protection of end users (patients, and persons with disabilities).

Considerable attention will be paid to experiences in Serbia as well as a comparative analysis of the use of VI in developed countries.

4 Conclusion

Artificial intelligence has great potential, not to solve but to participate in solving the most important social problems: ecology, climate change, economy, energy, and poverty reduction. However, it is clear that here, as well, a distinction is made between the rich and the poor, and that ethics and justice are called into question.

The application of AI in medicine and social services, in work with patients, i.e., users of social services and persons with disabilities, is a challenge for today's society. Society is faced with problems that must be solved in parallel with the technical progress and development of AI.

Any unwanted interaction between AI and people must not be allowed to occur. The responsibility and limits of the use of AI, especially in the decision-making segment, must be clearly defined in advance.

References

1. Andreeva, A., Dimitrova, D., Yolova, G.: Artificial intellect: regulatory framework and challenges facing the labour market. In: CompSysTech 2019: Proceedings of the 20th International Conference on Computer Systems and Technologies (2019)
2. Birhane, A., van Dijk, J.: Robot rights? Let's talk about human welfare instead. In: Proceedings of the AAAI/ACM Conference on AI, Ethics, and Society, pp. 207–213 (2020)
3. Branković, S.: Veštačka inteligencija i društvo. Srpska politička misao, broj 2/2017.god. Institut za političke studije, Beograd, 24. vol. 56, pp. 13–32 (2017)
4. von Braun, J., Archer, M.S., Reichberg, G.M., Sánchez Sorondo, M.: AI, robotics, and humanity: opportunities, risks, and implications for ethics and policy. In: von Braun, J., S. Archer, M., Reichberg, G.M., Sánchez Sorondo, M. (eds.) Robotics, AI, and Humanity, pp. 1–13. Springer, Cham (2021). https://doi.org/10.1007/978-3-030-54173-6_1

5. Dehaene, S., Lau, H., Kouider, S.: What is consciousness, and could machines have it? In: von Braun, J., S. Archer, M., Reichberg, G.M., Sánchez Sorondo, M. (eds.) Robotics, AI, and Humanity, pp. 43–56. Springer, Cham (2021). https://doi.org/10.1007/978-3-030-54173-6_4

6. De Backer, K., De Stefano, T.: Robotics and the global organization of production. In: von Braun, J., S. Archer, M., Reichberg, G.M., Sánchez Sorondo, M. (eds.) Robotics, AI, and Humanity, pp. 71–84. Springer, Cham (2021). https://doi.org/10.1007/978-3-030-54173-6_6

7. Fu Lee, K.: AI Super-Powers. China, Silicon Valley, and the New World Order. Haughton Mifflin Harcourt, Boston (2018). ISBN 9781328546395

8. Fu Lee, K., Qiufan, C.: AI 2041. Ten Vision For Our Future. Currency (2021). ISBN 9780593238295

9. Kamarinou, D., Millard, C., Singh J.: Machine Learning with Personal Data. Queen Mary University of London, School of Law Legal Studies, Research, Paper 247 (2016)

10. Ooi, V., Goh, G.: Taxation of automation and artificial intelligence as a tool of labour policy. eJournal Tax Res. **19**(2), 273–303 (2022)

11. Shum, H., He, X., Li, D.: From Eliza to XiaoIce: challenges and opportunities with social chatbots. Front. Inf. Technol. Electronic Eng. **19**, 10–26 (2018). https://doi.org/10.1631/FITEE.1700826

12. Singer, W.: Differences between natural and artificial cognitive systems. In: von Braun, J., S. Archer, M., Reichberg, G.M., Sánchez Sorondo, M. (eds.) Robotics, AI, and Humanity, pp. 17–27. Springer, Cham (2021). https://doi.org/10.1007/978-3-030-54173-6_2

13. Turing, A.M.: Computing machinery and intelligence (1950). The Essential Turing: the Ideas That Gave Birth to the Computer Age, pp. 433–464 (2012)

14. Zhang, C., Lu, Y.: Study on artificial intelligence: the state of the art and future prospects. J. Ind. Inf. Integr. **23**, 100224 (2021)

Advancing Bioprinting Technologies: PCL/PEG Polymers as Optimal Materials for 3D Scaffold Fabrication

Nebojsa Zdravković[1]([✉]), Sara Mijailović[1], Jelena Dimitrijević[1], Nikolina Kastratović[1], and Marko Živanović[2]

[1] Faculty of Medical Sciences, University of Kragujevac, Svetozara Markovića 69, 34000 Kragujevac, Serbia
nzdravkovic@gmail.com

[2] Institute for Information Technologies, University of Kragujevac, Jovana Cvijića Bb, 34000 Kragujevac, Serbia
marko.zivanovic@uni.kg.ac.rs

Abstract. In the realm of tissue engineering, 3D bioprinting has emerged as a cutting-edge methodology, unlocking unprecedented possibilities for the fabrication of tissue-like structures with potential applications in regenerative medicine. The advancement of 3D bioprinting techniques empowers the creation of intricate, organ-like structures that can either replace damaged portions of organs or serve as substitutes for entire organs. Notable successes in *in vivo* experiments involving 3D bioprinted skin, bone, and bladder underscore the transformative potential of this state-of-the-art technology.

The evolving landscape of healthcare underscores the imperative for a personalized therapeutic approach, necessitating innovative strategies in tissue engineering. The precision offered by modern techniques and devices, notably 3D bioprinters, enhances the success of this multidisciplinary scientific endeavor.

A primary focus of our research is the development of a method for the production of artificial blood vessels, responding to the high demands in the field of vascular treatment and healing. The crux of artificial blood vessel bioengineering lies in the utilization of meticulously crafted scaffolds, strategically seeded with stem cells that differentiate into somatic cells within human tissue. In pursuit of the optimal scaffold design for blood vessel production, we propose the application of polyethylene glycol (PEG) and polycaprolactone (PCL) polymers. Our results reveal a chemistry that proves to be optimal for this critical task, paving the way for advancements in the bioengineering of artificial blood vessels. This study contributes to the evolving landscape of tissue engineering and underscores the potential of PEG and PCL polymers in pioneering innovative solutions for vascular therapy.

Keywords: 3D Bioprinting · Scaffold Optimization · Tissue Engineering · Artificial Blood Vessel

N. Filipović (Ed.): AAI 2023, LNNS 999, pp. 194–200, 2024.
https://doi.org/10.1007/978-3-031-60840-7_24

1 Introduction

The fundamental objective of regenerative medicine and tissue engineering lies in the restoration, recovery, and replacement of afflicted or diseased tissue to reinstate normal physiological functions. This ambitious goal necessitates the deployment of a myriad of sophisticated methodologies, with bioprinting emerging as a dynamically evolving field at the forefront of these endeavors [1]. Bioprinting, characterized by its precise layer-by-layer application of biomaterials, leverages specialized software to construct three-dimensional structures with varying resolutions. The materials employed in bioprinting span a broad spectrum, encompassing natural substances such as fibrinogen, collagen, and gelatin, as well as chemically synthesized biocompatible materials like diverse polymers, among others. Often applied in conjunction with live cells, these materials create a conducive environment for the development of tissue-like constructs.

The versatility of bioprinting extends across multiple domains, encompassing tissue engineering, regenerative medicine, cancer investigation, transplantation, and drug development. Key milestones in tissue engineering achieved through bioprinting include the creation of bone-like structures, skin, and cardiac tissue structures [2–4].

Tissue engineering, as a discipline within this expansive landscape, predominantly involves the strategic use of highly porous biocompatible materials serving as scaffolds. These scaffolds play a pivotal role in providing a substrate for cell growth, and facilitating tissue regeneration or replacement [5]. To fulfill their purpose effectively, scaffolds must emulate the natural cellular environment, promoting normal growth, proliferation, and manifestation of physiological functions. Consequently, scaffold chemistry has emerged as a critical avenue of scientific exploration.

For a scaffold to be deemed suitable for regenerative medicine, it must satisfy three fundamental requirements: biocompatibility, biodegradability, and appropriate mechanical properties. In essence, the material should enable natural communication with cells, exhibit optimal mechanical properties, and undergo uniform biodegradation without toxic byproducts [6]. It is within this context that polyethylene glycol (PEG) and polycaprolactone (PCL) have gained prominence as widely employed materials.

This paper delves into the optimization of physical and chemical parameters pertaining to the utilization of PEG, PCL, and their combinations. The primary focus is on crafting scaffolds for tissue engineering, with a particular emphasis on fabricating structures mirroring the architecture of human blood vessels. The optimization of these polymers seeks to enhance the scaffolds' efficacy in fostering cell growth and tissue regeneration, especially in the intricate context of blood vessels.

The choice of PEG and PCL is grounded in their well-established attributes. PEG, with its hydrophilic nature, promotes biocompatibility and cellular interactions crucial for tissue engineering applications. PCL, prized for its biodegradability and mechanical strength, imparts structural integrity to the scaffold. The synergy between these polymers presents a promising avenue for advancing the field of tissue engineering, especially in the challenging domain of mimicking the complexity of blood vessels.

As we delve into the intricacies of optimizing these polymers, the overarching goal is to contribute to the refinement of scaffold design, pushing the boundaries of what is achievable in tissue engineering. This exploration holds the potential to not only deepen our understanding of polymer behavior in a biological context but also pave the way for innovative solutions in regenerative medicine, particularly in the quest to engineer functional human blood vessels.

2   Material, Methods and Results

The experimental framework of this study revolves around the utilization of the layer-by-layer 3D bioprinting method to fabricate scaffolds with optimized physical and chemical properties. Our primary focus involves exploring a wide range of PCL and PEG solutions, both individually and in combination, to discern the most effective parameters for scaffold design. The sophisticated 3D bioprinting technology at the core of this study is facilitated by the Inkredible+ device (CellInk, Gothenburg, Sweden), a cutting-edge bioprinter equipped with dual printheads and a UV LED system essential for crosslinking bioprints (Fig. 1).

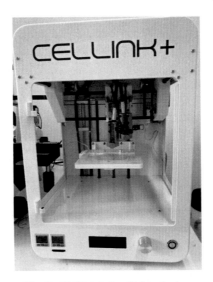

Fig. 1. CellInk Inkredible+ device.

Bioprinting System:
The Inkredible+ 3D bioprinter, being the workhorse of our experimental setup, is intricately connected to a computer and supported by specialized software that meticulously controls the units and the entire bioprinting process. The device relies on three-dimensional CAD models of scaffolds, which are translated into precise coordinates, laying the foundation for the layer-by-layer application of polymers. The bioprinting

process operates on a pressure extrusion mechanism, allowing controlled polymer dispensing to construct complex three-dimensional structures. This mechanism is finely tuned, with an XY position precision of 10 μm, Z precision of 2.5 μm, and a layer resolution of 100 μm. To contain the bioprints during fabrication, all samples are systematically printed in a P100 Petri dish.

Materials and Solvents:
Our exploration into scaffold optimization begins with the careful consideration of materials and solvents. The selection of PCL and PEG, both widely recognized for their biocompatibility, mechanical strength, and hydrophilic nature, lays the foundation for a synergistic approach to scaffold design. The solvent system, chloroform/DMF, serves as a crucial determinant, balancing solubility and evaporation rates to ensure precise layering during the bioprinting process.

Parameter Optimization:
A critical phase of our study involves the systematic optimization of various parameters. The PCL and PEG ratio combinations are meticulously varied to discern their impact on scaffold properties. PCL, known for its biodegradability and mechanical strength, is balanced against PEG, chosen for its hydrophilic nature enhancing biocompatibility. The ratios are systematically adjusted, creating a spectrum of scaffolds, each potentially offering unique advantages in terms of structural integrity and cell interactions.

Temperature modulation emerges as a key variable influencing the physical characteristics of the bioprinted scaffolds. By systematically altering the applied temperature, we aim to influence the rate of solvent evaporation and polymer crosslinking, thereby tailoring the structural features of the scaffolds. This parameter becomes a key determinant in achieving the desired balance between porosity, mechanical strength, and biodegradability.

Bioprinting Process:
The bioprinting process is meticulously orchestrated, beginning with the translation of CAD models into precise coordinates. This process dictates the intricate layering of polymers during the printing process. Pressure extrusion is employed, enabling controlled flow and deposition of the polymer to create complex structures. The dual printheads of the Inkredible+ device facilitate the simultaneous deposition of multiple materials, allowing the creation of multifunctional scaffolds.

The UV LED system integrated into the bioprinting system plays a critical role in the cross-linking of the bioprinted materials. This step is pivotal for creating scaffolds capable of withstanding subsequent cell seeding and in vivo applications. The entire process, from the translation of digital models to the crosslinking of bioprints, is executed with precision, validating the capabilities of the Inkredible+ 3D bioprinter.

Scaffold Characterization and Further Examinations:
The optimization process yields a diverse series of scaffolds, each poised for subsequent examination. Our focus now shifts to assessing the impact of these scaffolds on cell seeding, evaluating their influence on cell viability, and understanding their biodegradability profiles. These critical analyses will provide invaluable insights into the functional efficacy of the scaffolds, setting the stage for their potential application in tissue engineering, with a primary emphasis on the production of human blood vessel structures.

Biocompatibility and Cell Seeding:

The biocompatibility of scaffolds is a critical aspect that directly influences their efficacy in tissue engineering. Our next phase involves the thorough examination of how cells interact with the fabricated scaffolds. Cell seeding experiments will be conducted to evaluate the ability of the scaffolds to support cell attachment, proliferation, and differentiation. Fluorescent markers will be employed to visualize and quantify these cellular responses, providing a comprehensive understanding of the biocompatibility of the scaffolds.

Influence on Cell Viability:

Understanding the impact of scaffolds on cell viability is essential for predicting their success in regenerative medicine applications. A battery of assays, including live/dead staining and metabolic activity assessments, will be employed to evaluate the influence of the scaffolds on the health and viability of the seeded cells. This thorough analysis will guide us in refining the scaffold design to enhance cell survival and functionality.

Biodegradability Profiles:

The biodegradability of scaffolds is a critical determinant of their suitability for in vivo applications. We will assess the degradation kinetics of the scaffolds under controlled conditions, monitoring changes in mass, morphology, and mechanical properties over time. Understanding the biodegradability profiles will aid in tailoring the scaffolds to match the dynamic regeneration process within the human body, ensuring a harmonious integration of the engineered tissues (Fig. 2 and Table 1).

Fig. 2. Scaffold geometry for 3D bioprint.

Table 1. Solutions and parameters summary

Solvent	Dielectric constant at 20° (ε)	Electrical conductivity ($S.m^{-1}$)	Surface Tension ($mN.m^{-1}$)	Boiling temp (°C)
Methanol	6.7	1.5×10^{-7}	22.6	64.7
Chloroform	4.8	$<1 \times 10^{-8}$	27.16	61.15
DMF	36.7	6×10^{-6}	35	152

3 Conclusions

The pursuit of optimizing various experimental parameters for the creation of scaffolds tailored for tissue engineering is an intricate and demanding process. As we navigate through this multifaceted endeavor, our ultimate goal is to engineer scaffolds with meticulously designed properties that will serve as a foundational substrate for the seeding of endothelial and smooth muscle cells. This intricate process aims to unravel the suitability of the fabricated scaffolds for the ambitious task of creating an artificial blood vessel in vitro.

The significance of a well-optimized scaffold cannot be overstated, as it forms the crux of our strategy towards developing an artificial blood vessel. The subsequent steps involve the strategic seeding of endothelial and smooth muscle cells onto these scaffolds, simulating the physiological conditions necessary for the formation of a functional blood vessel. This pivotal phase acts as a bridge between the laboratory-based experimentation and the transformative potential of applying these constructs within the complex landscape of the human body.

The selection of a suitable scaffold marks the commencement of a transformative journey toward the creation of a functional blood vessel in vitro. The successful attachment of cells to the scaffold represents a crucial milestone in the development of a novel model that aspires to function as a blood vessel within the body. However, the realization of this ambitious objective hinges on the successful completion of subsequent animal and clinical trials. These in vivo testing stages are pivotal in determining the scaffold's rate of biodegradation and the nature of the immune response elicited by the implantation.

The intricate interplay between scaffold materials and their impact on degradation rates and immune responses underscores the complexity of this phase in the development process. The selection of materials profoundly influences the scaffold's integration within the biological milieu, and careful consideration is paramount in ensuring not only the structural integrity of the scaffold but also its biocompatibility and functionality.

In the context of developing an artificial blood vessel, the stakes are particularly high during the initial stages. The importance of meticulous attention to detail in scaffold formation cannot be overstated. The scaffolds must not only be structurally sound but also possess the desired properties conducive to cellular attachment, proliferation, and functionality. As we navigate this critical juncture in the development of tissue-engineered blood vessels, the focus remains on refining the scaffolds to meet the stringent criteria required for successful in vivo applications.

Beyond the controlled environment of the laboratory, the scaffolds must seamlessly integrate with the dynamic and complex environment of the human body. The rate of scaffold biodegradation and the nature of the immune response become pivotal factors in determining the long-term success and sustainability of the tissue-engineered blood vessel. Consequently, the insights gained from animal and clinical trials will be instrumental in informing the iterative refinement of scaffold design and composition.

The bridge between in vitro success and in vivo applicability relies on a nuanced understanding of the physiological intricacies associated with the vascular system. The artificial blood vessel, once developed, holds the promise of revolutionizing regenerative medicine and addressing critical challenges in vascular therapy. However, this journey is a continuum, demanding unwavering dedication to scientific rigor, ethical considerations, and the pursuit of excellence in biomaterial design.

In conclusion, the optimization of experimental parameters marks an essential phase in the intricate process of developing tissue-engineered blood vessels. The carefully designed scaffolds, poised for cellular seeding, represent a critical advancement toward realizing the vision of creating an artificial blood vessel in vitro. The journey ahead, encompassing animal and clinical trials, demands a meticulous and thorough exploration of scaffold performance in vivo. As we navigate through these crucial stages, the insights gained will not only shape the future trajectory of tissue engineering but also hold the potential to redefine the landscape of regenerative medicine. The pursuit of engineering functional blood vessels is an embodiment of scientific ambition, and our commitment to excellence in scaffold design is an unwavering testament to the transformative potential of biotechnology in the service of human health and well-being.

Acknowledgement. The research was funded by the Ministry of Science, Technological Development and Innovation of the Republic of Serbia, contract number [451-03-47/2023-01/200107 (Faculty of Engineering, University of Kragujevac), 451-03-47/2023-01/200378 (Institute for Information Technologies Kragujevac, University of Kragujevac)]. This work is supported by the European Union's Horizon 2020 research and innovation programme under grant agreement No 952603 (SGABU). This article reflects only the author's view. The Commission is not responsible for any use that may be made of the information it contains.

References

1. Ozbolat, I.T.: Bioprinting scale-up tissue and organ constructs for transplantation. Trends Biotechnol. **33**, 395–400 (2015)
2. Bose, S., Vahabzadeh, S., Bandyopadhyay, A.: Bone tissue engineering using 3D printing. Mater. Today **16**(12), 496–504 (2013)
3. Augustine, R.: Skin bioprinting: a novel approach for creating artificial skin from synthetic and natural building blocks. Prog. Biomater. **7**(2), 77–92 (2018)
4. Qasim, M., Haq, F., Kang, M.H., Kim, J.H.: 3D printing approaches for cardiac tissue engineering and role of immune modulation in tissue regeneration. Int. J. Nanomed. **20**(14), 1311–1333 (2019)
5. Pati, F., Gantelius, J., Svahn, H.A.: 3D bioprinting of tissue/organ models. Angew. Chem. Int. Ed. **55**(15), 4650–4665 (2016)
6. Brennan, A.B., Kirschner, C.M.: Bio-inspired Materials for Biomedical Engineering. Wiley, New York (2014)

Simulation of Blood Flow Through a Patient-Specific Carotid Bifurcation Reconstructed Using Deep Learning Based Segmentation of Ultrasound Images

Tijana Djukic[1,2]($\boxtimes$), Milos Anic[2,3], Branko Gakovic[4], Smiljana Tomasevic[2,3], Branko Arsic[2,5], Igor Koncar[4], and Nenad Filipović[2,3]

[1] Institute for Information Technologies, University of Kragujevac, Jovana Cvijica bb, 34000 Kragujevac, Serbia
tijana@kg.ac.rs

[2] Bioengineering Research and Development Center, BioIRC, Prvoslava Stojanovica 6, 34000 Kragujevac, Serbia
fica@kg.ac.rs

[3] Faculty of Engineering, University of Kragujevac, Sestre Janjica 6, 34000 Kragujevac, Serbia

[4] Clinic for Vascular and Endovascular Surgery, Serbian Clinical Centre, Dr Koste Todorovica 8, 11000 Belgrade, Serbia

[5] Faculty of Science, University of Kragujevac, Radoja Domanovica 12, 34000 Kragujevac, Serbia

Abstract. One of the diseases of the cardiovascular system is the formation of carotid artery stenosis. The existence of atherosclerotic plaque within the vessel wall causes changes in blood flow and can have serious consequences to the individual's health condition. Therefore early and appropriate clinical diagnostics is very important. One of the first clinical examinations for this disease is the ultrasound (US) examination. Three-dimensional (3D) reconstruction and blood flow simulation could be used to overcome some of the drawbacks of the US examination and improve the overall diagnostics. An approach that combines the deep learning techniques and 3D reconstruction and meshing algorithms is applied within this study to first create the model of patient-specific carotid bifurcation and then to perform unsteady blood flow simulation, with realistic boundary conditions. This type of simulations can provide quantitative hemodynamic data to the clinicians during US examination and can further help to improve the diagnostics and ensure a treatment that is more adapted to the particular patient.

Keywords: deep learning · image segmentation · 3D reconstruction · finite element method · unsteady blood flow

1 Introduction

One of the diseases of the cardiovascular system is the formation of carotid artery stenosis. The existence of atherosclerotic plaque within the vessel wall causes changes in blood flow. This can lead to the reduction of blood supplies to the brain, and ultimately

N. Filipović (Ed.): AAI 2023, LNNS 999, pp. 201–206, 2024.
https://doi.org/10.1007/978-3-031-60840-7_25

to serious ischemic consequences. Therefore, early and appropriate clinical diagnostics is very important. One of the first clinical examinations for this disease is the ultrasound (US) examination. However, the amount of information obtained from US images is rather limited. Only two-dimensional (2D) cross-sections are obtained which limits the overview of the entire geometry of the carotid bifurcation, and the information about the hemodynamics within the entire carotid bifurcation are not available. Deep learning represents a useful tool that can enable more in-depth analysis of available clinical images [1, 2]. The numerical simulations of blood flow can provide useful quantitative information about the state of the blood vessels, like it was already demonstrated in literature [3, 4]. The approach that could be used to overcome the mentioned drawbacks of the US examination and improve the overall diagnostics is a three-dimensional reconstruction and blood flow simulation. A solution was proposed in literature [5] that combines the deep learning techniques and 3D reconstruction and meshing techniques to create the model of patient-specific carotid bifurcation. This approach is applied within this study to perform unsteady blood flow simulation, with realistic boundary conditions and analyze the hemodynamic parameters.

The paper is organized as follows. Details of the reconstruction methodology and numerical method used for blood flow simulation are discussed in Sect. 2. Results for the unsteady 3D blood flow are presented in Sect. 3. Section 4 concludes the paper.

2   Materials and Methods

The applied 3D reconstruction technique is presented in Sect. 2.1, while the numerical model used for the blood flow simulation is presented in Sect. 2.2.

2.1   3D Reconstruction

The applied methodology for the 3D reconstruction that used deep learning segmentation of US images is described extensively in literature [5–7]. The U-Net [8] based deep convolutional neural networks (CNN) is used for the segmentation purpose. First a clinical data set containing patient-specific US images is used to train the neural network. Within the data set, both longitudinal and transversal US images are present and hence two CNNs were trained independently. These CNNs are then used with previously unknown US images to segment the regions of lumen and arterial wall for the specific patient. Since the main goal of this study is to perform blood flow simulations, only the lumen areas were of interest and only these regions were reconstructed. The longitudinal US images are used to define the centerlines of the branches of carotid bifurcation and the transversal US images are used to define the cross-sections of the branches. This information is used to create a 3D geometry that consists of hexahedral finite elements and this geometry is then used to perform the blood flow simulation.

2.2 Blood Flow Simulation

In order to simulate the unsteady 3D blood flow, the Navier-Stokes equations and continuity equation are used:

$$\rho\left(\frac{\partial v_i}{\partial t} + v_j\frac{\partial v_i}{\partial x_j}\right) = -\frac{\partial p}{\partial x_i} + \mu\left(\frac{\partial^2 v_i}{\partial x_j \partial x_j} + \frac{\partial^2 v_j}{\partial x_j \partial x_i}\right), \tag{1}$$

$$\frac{\partial v_i}{\partial x_i} = 0, \tag{2}$$

In Eqs. (1) and (2), v_i denotes the component of the velocity in direction x_i, p is the blood pressure and the blood characteristics - density and dynamic viscosity are denoted by ρ and μ, respectively.

Another important hemodynamic quantity, besides velocity and pressure is the wall shear stress (WSS). This quantity can be calculated within the simulation based on the tangential velocity $\mathbf{u}_t$ using the following equation:

$$\tau = -\mu\frac{\partial \mathbf{u}_t}{\partial \mathbf{n}}, \tag{3}$$

where $\mathbf{n}$ denotes the unit vector that is normal to the vessel wall in the considered point of the mesh.

Since in this study, an unsteady blood flow simulation will be considered, another quantity related to WSS will be also calculated – the time averaged WSS (TAWSS) that is given by:

$$TAWSS = \frac{1}{T}\int_0^T |\tau|dt, \tag{4}$$

Equations (1) and (2) are transformed into an incremental-iterative form and they are solved in iterations using the finite element method [9]. The simulations are performed within the in-house developed software PakF [3, 10]. The convergence criterion is defined such that the absolute change in the non-dimensional velocity within the current time step should be less than 10^{-3}.

3 Results

Within this study, the US images of one patient are used to perform the patient-specific 3D reconstruction and blood flow simulation. The reconstructed geometry is shown in Fig. 1. Within the simulation, the following boundary and initial conditions in simulations are defined. It is considered that the arterial wall is rigid and all nodes of the arterial wall have zero velocity. At the outlet of the internal carotid artery (ICA), the outflow zero pressure boundary condition is applied. At the inlet of the common carotid artery (CCA) and at the outlet of the external carotid artery (ECA), the time-dependent parabolic velocity profiles are prescribed according to the measured blood flow for the specific patient. In Fig. 1, the prescribed change of velocity over time for both inlet of the CCA and outlet of the ECA are shown above the reconstructed geometry for this particular case. Overall simulation time was equal to one cardiac cycle, i.e. 0.8 s.

The results of the blood flow simulation are shown in Fig. 2. Two characteristic time points were chosen for the presentation of results - peak systole (t = 0.2 s) and peak diastole (t = 0.4 s). The pressure distribution is shown in Figs. 2A and 2B, the velocity streamlines are shown in Figs. 2C and 2D and the WSS distribution is shown in Figs. 2E and 2F. The time averaged distribution of WSS for the whole simulation is shown in Fig. 3A, while in Fig. 3B the parts of the artery with TAWSS lower than 1 MPa are denoted in blue and higher than 2.5 MPa are denoted in red.

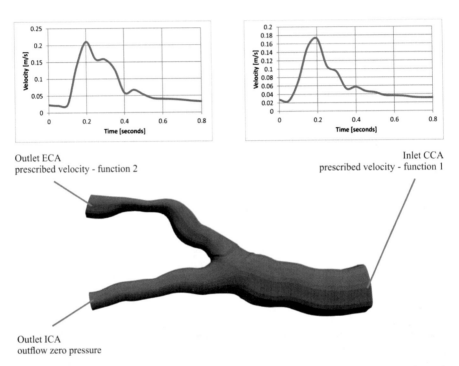

Fig. 1. Results of 3D reconstruction and blood flow simulation for a particular patient; A – Reconstructed 3D finite element mesh; B – Pressure distribution at peak systole; C – Velocity streamlines at peak diastole; D - time averaged distribution of WSS.

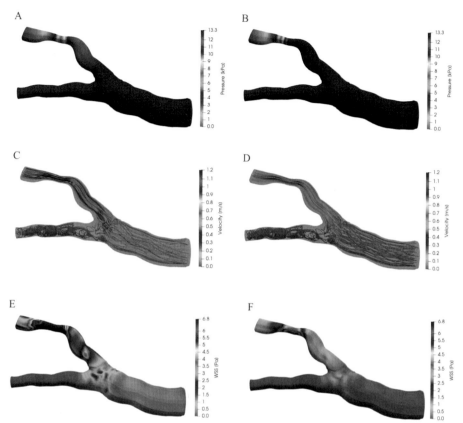

Fig. 2. Results of blood flow simulation for a particular patient; A – Pressure distribution at peak systole; B – Pressure distribution at peak diastole; C – Velocity streamlines at systole diastole; D – Velocity streamlines at peak diastole; E – WSS distribution at peak systole; F – WSS distribution at peak diastole.

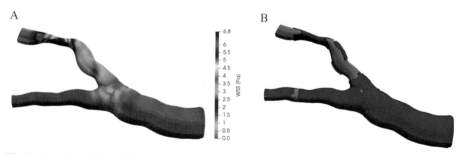

Fig. 3. Results of blood flow simulation for a particular patient; A – TAWSS distribution; B – Extracted parts of the carotid bifurcation with high TAWSS (over 2.5 MPa, in red) and low TAWSS (below 1 MPa, in blue). (Color figure online)

4 Conclusions

The approach that combines deep learning techniques with 3D reconstruction and meshing algorithms enabled an efficient segmentation of US images of patient-specific carotid bifurcation and creation of 3D mesh that was further used for the numerical simulation of blood flow. This type of simulations can provide quantitative data to the clinicians during US examination, above all information that cannot be easily measured, such as the distribution of WSS. This information can further help improve the diagnostics and ensure a treatment that is more adapted to the particular patient.

Acknowledgment. The research presented in this study was part of the project that has received funding from the European Union's Horizon 2020 research and innovation programme under grant agreement No. 755320-2 - TAXINOMISIS. This article reflects only the author's view. The Commission is not responsible for any use that may be made of the information it contains.

References

1. Shen, D., Wu, G., Suk, H.I.: Deep learning in medical image analysis. Annu. Rev. Biomed. Eng. **19**, 221–248 (2017)
2. Ravì, D., et al.: Deep learning for health informatics. IEEE J. Biomed. Health Inform. **21**(1), 4–21 (2016)
3. Parodi, O., et al.: Patient-specific prediction of coronary plaque growth from CTA angiography: a multiscale model for plaque formation and progression. IEEE Trans. Inf. Technol. Biomed. **16**(5), 952–965 (2012). https://doi.org/10.1109/TITB.2012.2201732
4. Djukic, T., Topalovic, M., Filipovic, N.: Validation of lattice Boltzmann based software for blood flow simulations in complex patient-specific arteries against traditional CFD methods. Math. Comput. Simul. **203**, 957–976 (2023)
5. Anic, M., Djukic, T., Gakovic, B., Arsic, B., Filipovic, N.: Improved three-dimensional reconstruction of patient-specific carotid bifurcation using deep learning based segmentation of ultrasound images. In: 1st Serbian International Conference on Applied Artificial Intelligence (SICAAI), Kragujevac, Serbia, 19–20 May 2022
6. Djukic, T., Arsic, B., Koncar, I., Filipovic, N.: 3D reconstruction of patient-specific carotid artery geometry using clinical ultrasound imaging. In: Miller, K., Wittek, A., Nash, M., Nielsen, P.M.F. (eds). Computational Biomechanics for Medicine, pp. 73–83. Springer, Cham (2020). https://doi.org/10.1007/978-3-030-70123-9_6
7. Djukic, T., Arsic, B., Djorovic, S., Koncar, I., Filipovic, N.: Validation of the machine learning approach for 3D reconstruction of carotid artery from ultrasound imaging. In: IEEE 20th International Conference on Bioinformatics and Bioengineering (BIBE), 26–28 October 2020
8. Ronneberger, O., Fischer, P., Brox, T.: U-net: convolutional networks for biomedical image segmentation. In: Navab, N., Hornegger, J., Wells, W.M., Frangi, A.F. (eds.) MICCAI 2015, pp. 234–241. Springer, Cham (2015). https://doi.org/10.1007/978-3-319-24574-4_28
9. Kojic, M., Filipovic, N., Stojanovic, B., Kojic, N.: Computer Modeling in Bioengineering: Theoretical Background, Examples and Software. Wiley, Chichester (2008)
10. Filipovic, N., Mijailovic, S., Tsuda, A., Kojic, M.: An implicit algorithm within the Arbitrary Lagrangian-Eulerian formulation for solving incompressible fluid flow with large boundary motions. Comp. Meth. Appl. Mech. Engrg. **195**, 6347–6361 (2006)

Feature Selection for the Shear Stress Classification of Hip Implant Surface Topographies

Aleksandra Vulović[1,2]([✉]), Tijana Geroski[1,2], and Nenad Filipović[1,2]

[1] Faculty of Engineering, University of Kragujevac, Sestre Janjić 6, 34000 Kragujevac, Serbia
{aleksandra.vulovic,tijanas,fica}@kg.ac.rs
[2] Bioengineering Research and Development Center (BioIRC), Prvoslava Stojanovića 6, 34000 Kragujevac, Serbia

Abstract. In order to determine the optimal hip implant surface topography that minimizes the shear stress during everyday activities, a large number of models need to be created and analyzed. To understand how different model parameters affect the shear stress values and distributions, the parameters are varied during the model creation process. Depending on the complexity of the model and the number of elements for the finite element simulation, the time needed to obtain the results can vary from a few minutes to a few hours. The aim of this study was to analyze the application of feature selection algorithms as a way to understand which parameters have the highest impact on the shear stress results. Two algorithms were considered – a tree-based model and Principal Component Analysis. The used dataset consisted of 64 models, previously analyzed using the finite element method. There were 11 input parameters and a single target variable. This approach can reduce the number of numerical models that need to be created and analyzed, thus saving time and resources. The obtained results indicated that the most important parameters can be extracted even when working with a small dataset such as the one considered in this paper.

Keywords: feature selection · hip implant · surface topographies

1 Introduction

The process of numerical analysis of any model includes a number of steps such as defining a model geometry and creating a model of implant and femoral bone, automatically or manually creating a mesh, defining material properties and appropriate boundary conditions, and running simulations [1]. The process of model analysis can take from a few minutes to a few hours depending on the model's complexity and the number of elements. The process of defining the optimal hip implant surface topography requires analysis of a large number of parameters and their varieties which requires a lot of time.

Due to the high time demand of such a study, it is essential to find new methods to reduce the number of models that are created and numerically analyzed. Therefore, the application of machine learning algorithms was considered as a way to decrease the

N. Filipović (Ed.): AAI 2023, LNNS 999, pp. 207–213, 2024.
https://doi.org/10.1007/978-3-031-60840-7_26

number of different parameters that are analyzed and the number of models that are developed [2]. After the application of classification algorithms for defining the combination of model parameters, the possibility of extracting the most significant parameters was considered.

Machine learning algorithms have also been widely applied to other problems in the field of orthopedics, besides hip implants. Various methods have been used to detect joint and bone fractures [3–5], osteoporosis [6, 7], and osteoarthritis [8, 9]. Machine learning algorithms have been used for the early recognition of a failed total hip joint replacement procedure [10] as well as for assessing the risk of hip dislocation after total hip replacement [11].

The aim of this paper was to analyze if feature selection is a possible option for reducing the number of potential models that should be developed and analyzed. It is difficult to reach conclusions about the importance of those parameters based on the results of numerical simulations without the help of algorithms that enable attribute extraction. In this case, two methods were considered: a model based on a decision tree and analysis of principal components - Principal Component Analysis - PCA.

2   Materials and Methods

For the feature selection considered in this paper, two approaches have been considered - training a tree-based model and Principal Component Analysis. There were 11 input parameters and a single target variable (maximum shear stress value). The table contained 64 rows and 12 columns, where column 12 was the target variable. The list of the considered parameters is given in Table 1.

Table 1. List of considered parameters.

Number	Parameter	Type	Value
1	Number of half-cylinders lengthwise	Integer	>1
2	Number of half-cylinder rows	Integer	>0
3	Half cylinder added or removed from the surface	Integer	0 or 1
4	Distance between half cylinders lengthwise	Real	≥0
5	Distance between half cylinders widthwise	Real	≥0
6	Number of different radius values	Integer	1 or 2
7	Radius 1 value	Real	>0
8	Radius 2 value	Real	≥0
9	Distance from the edge where loading is located	Real	≥0
10	Distance from the other edge of the model	Real	≥0
11	Model includes trabecular bone	Integer	0 or 1

The list of parameters was similar to the list used in our previous study [2]. Half-cylinder orientation was no longer considered an important parameter. Instead, two new parameters were considered: if a half-cylinder was added or removed from the implant surface and if a trabecular bone model was created.

In the first method, a tree-based model, gradient augmentation was applied, which facilitates obtaining significance value for each considered parameter. The result of this method is obtaining a rating that indicates the relative importance of each parameter, i.e. whether the parameter is used more or less when making decisions. The obtained values provide us with the possibility to compare the ratings of the parameters and rank them.

The second method, PCA, is a statistical procedure used to reduce the dimensionality of the input data. This method is based on the application of orthogonal transformation, which transforms the input parameters, between which there may be a correlation, into parameters between which there is no correlation. These new parameters are called principal components, and their maximum number can be less than or equal to the number of original parameters. The first main component has the highest variance value, and it is possible to use it in order to describe the highest variability of the analyzed data.

The analysis was performed using a Jupyter notebook, where the *xgboost* [12] and *sklearn* [13] libraries were loaded with which the analysis was performed. For the tree-based model, the *xgboost* library was used, while the *sklearn* library was used for the principal component analysis.

The data was divided into two sets – a training set and a testing set. The training set contained 60% of the data, while the test set contained 40%. Due to having a relatively small database, cross-validation of data was used. In this case, 5-fold cross-validation was used.

It is important to note that the application of machine learning algorithms was restricted to data from previously performed analyses, which means that the data set used was small.

3   Results and Discussion

Tree-Based Model
Table 2 presents the results of the model evaluation. All four metrics were above 80%.

Table 2. Model evaluation results.

Accuracy	81%
Precision	84%
Recall	81%
F1 - score	81%

Individual values for each parameter (Table 1) as well as the parameter ranks are shown in Table 3. The total sum of all parameters was equal to 1.

It is noticeable that 3 parameters stand out - Parameter 2 (Number of semi-cylinders per width), Parameter 3 (Semi-cylinder added or removed from the surface), and Parameter 6 (Number of different diameter values). The results indicated that these three

Table 3. Parameter values.

Rank	Parameter	Value
1	Parameter 3	0.289
2	Parameter 6	0.223
3	Parameter 2	0.215
4	Parameter 4	0.091
5	Parameter 7	0.069
6	Parameter 9	0.046
7	Parameter 1	0.042
8	Parameter 5	0.018
9	Parameter 10	0.007
10	Parameter 8	0
11	Parameter 11	0

parameters have the greatest influence on the value of the numerical simulation, in terms of whether the shear stress on the implant surface will be lower or higher than 10 MPa.

Although the accuracy and precision of the results were slightly higher than 80%, it was still possible to reach important conclusions regarding parameters that showed the greatest importance.

Principal Component Analysis

The results of the principal components analysis are shown in Fig. 1. The obtained results showed that three principal components are sufficient to describe more than 95% of the variances. Table 4 shows the numerical results of the connection of the first 3 main components with 11 input parameters (Table 1). The first principal component is the most significant because it can explain over 90% of the variance.

The first principal component was highly correlated with two parameters - Parameter 4 (Distance between half-cylinders in length) and Parameter 7 (Diameter value 1). Based on the correlation values, it can be concluded that the main component was primarily related to the distance between half-cylinders along the length. The second principal component was also highly correlated with the same two parameters as the first principal component. In this situation, an increase in one parameter did not lead to an increase in the other, instead, it led to a decrease. Unlike the first principal component, the second principal component was primarily associated with the diameter value of 1. The third principal component was highly correlated with parameters 9 (Distance of the first half-cylinder from the end where the load is defined) and 10 (Distance of the last half-cylinder from the other end of the model). This component increased with the increase of these two parameters. To the best of our knowledge, no similar research has been performed so far, and thus there are no results that can be compared.

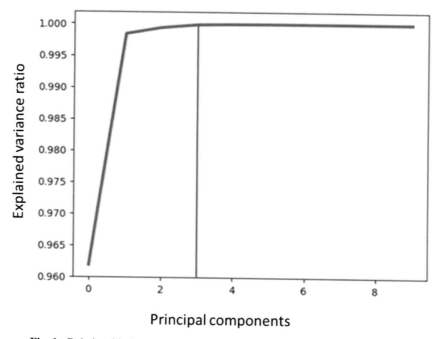

Fig. 1. Relationship between the number of principal components and variance.

Table 4. Results of principal component analysis

Parameter	Principal component 1	Principal component 2	Principal component 3
Parameter 1	0.00182	0.0194	−0.186
Parameter 2	−0.000654	−0.00109	−0.00619
Parameter 3	0.000568	0.000221	−0.00392
Parameter 4	**0.916**	**−0.402**	−0.00483
Parameter 5	−0.01	−0.0178	0.0017
Parameter 6	−0.000176	−0.000388	−0.00887
Parameter 7	**0.402**	**0.915**	0.0233
Parameter 8	−0.000563	−0.00127	−0.0284
Parameter 9	−0.00615	−0.00854	**0.742**
Parameter 10	−0.0000311	−0.0207	**0.643**
Parameter 11	0.000937	0.00169	−0.00643

4 Conclusions

This paper presented the application of machine learning algorithms for providing additional information regarding models of modified hip implant surfaces. The paper used two types of algorithms: one for extracting the most significant parameters that affect the shear stress distribution on the hip implant surfaces, and the other for classifying the models based on their numerical results.

The obtained results for extracting the most significant parameters and classifying the models showed that the application of machine learning algorithms can be useful for providing additional information regarding the modified hip implant surfaces. This approach significantly shortens the time, because it indicates the most important parameters that influence the final values of the numerical analysis as well as the combination of parameters for which the models should be created for the numerical simulation.

The main limitation is related to the amount of data from previously performed finite element analysis, which has an impact on the obtained results. However, the obtained results show that machine learning algorithms can be used as a valuable tool for analyzing complex problems in biomechanics.

Acknowledgment. This research is supported by the European Union's Horizon 2020 research and innovation programme under grant agreement No. 952603 - SGABU. This article reflects only the author's The Commission is not responsible for any use that may be made of the information it contains. Authors also acknowledge the funding by the Ministry of Science, Technological Development and Innovation of the Republic of Serbia, contract number [451-03-47/2023-01/200107 (Faculty of Engineering, University of Kragujevac)].

References

1. Vulović, A., Filipovic, N.: Computational analysis of hip implant surfaces. J. Serb. Soc. Comput. Mech. **13**(1), 109–119 (2019)
2. Vulović, A., Šušteršič, T., Filipović, N.: Shear stress classification for the finite element analysis of hip implant surface topographies. In: 1st Serbian International Conference on Applied Artificial Intelligence (SICAAI) Kragujevac, Serbia, 19–20 May 2022, p. 88 (2022)
3. Langerhuizen, D.W., et al.: What are the applications and limitations of artificial intelligence for fracture detection and classification in orthopaedic trauma imaging? A systematic review. Clin. Orthop. Relat. Res. **477**(11), 2482 (2019)
4. Scala, A., Borrelli, A., Improta, G.: Predictive analysis of lower limb fractures in the orthopedic complex operative unit using artificial intelligence: the case study of AOU Ruggi. Sci. Rep. **12**(1), 22153 (2022)
5. Niiya, A., et al.: Development of an artificial intelligence-assisted computed tomography diagnosis technology for rib fracture and evaluation of its clinical usefulness. Sci. Rep. **12**(1), 8363 (2022)
6. Gao, L., Jiao, T., Feng, Q., Wang, W.J.: Application of artificial intelligence in diagnosis of osteoporosis using medical images: a systematic review and meta-analysis. Osteoporos. Int. **32**, 1279–1286 (2021)
7. Ferizi, U., et al.: Artificial intelligence applied to osteoporosis: a performance comparison of machine learning algorithms in predicting fragility fractures from MRI dana. J. Magn. Reson. Imaging **49**(4), 1029–1038 (2019)

8. Lee, L.S., et al.: Artificial intelligence in diagnosis of knee osteoarthritis and prediction of arthroplasty outcomes: a review. Arthroplasty **4**(1), 16 (2022)
9. Lee, K.S., et al.: Automated detection of TMJ osteoarthritis based on artificial intelligence. J. Dent. Res. **99**(12), 1363–1367 (2020)
10. Loppini, M., Gambaro, F.M., Chiappetta, K., Grappiolo, G., Bianchi, A.M., Corino, V.D.: Automatic identification of failure in hip replacement: an artificial intelligence approach. Bioengineering **9**(7), 288 (2022)
11. Rouzrokh, P., et al.: Deep learning artificial intelligence model for assessment of hip dislocation risk following primary total hip arthroplasty from postoperative radiographs. J. Arthroplasty **36**(6), 2197–2203 (2021)
12. Chen, T., Guestrin, C.: Xgboost: a scalable tree boosting system. In: Proceedings of the 22nd ACM SIGKDD International Conference on Knowledge Discovery and Data Mining, pp. 785–794 (2016)
13. Pedregosa, F., et al.: Scikit-learn: machine learning in Python. J. Mach. Learn. Res. **12**, 2825–2830 (2011)

Embedding Artificial Intelligence into Wearable IoMT Systems

Steven Puckett[1,2]([✉]), Vineetha Menon[3], and Emil Jovanov[1]

[1] Electrical and Computer Engineering, University of Alabama at
Huntsville, Huntsville, AL, USA
emil.jovanov@uah.edu
[2] Sanders College of Business and Technology, University of North Alabama,
Florence, AL, USA
spuckett1@una.edu
[3] Department of Computer Science, University of Alabama at Huntsville, Huntsville, AL, USA
vineetha.menon@uah.edu

Abstract. Recent advances in sensors, wearable technologies, and real-time embedded Artificial Intelligence (AI) allows the creation of wearable systems for continuous health monitoring and real-time detection of critical health issues. In the case of heart arrhythmias, this currently requires data to be transmitted to the cloud for processing by computationally heavy models and human interaction. Instead, real-time analysis by AI models on the wearable device would reduce communication transmissions, improve energy usage, and provide immediate feedback to the user and physicians. We have shown that AI models trained using Big Data datasets on larger platforms can be optimized and then deployed as wearable sensors integrated into Internet of Medical Things (IoMT) systems. We used Artificial Intelligence models to detect abnormal heart beats trained using the MIT-BIH dataset. Our objective is to identify heartbeat abnormalities using a model optimized for embedded wearable sensors. The system would warn the user in the case of abnormal heart beats and upload all detected abnormal beats to the medical server for evaluation by a cardiologist. The system was 98.4% accurate in classification of normal/abnormal heartbeats and only missed approximately 0.13% arrhythmias per day with an execution time per heart beat classification of 65.6 ms and uses 0.911 microwatt-hours of energy. The program requires 409 KB of additional memory of ESP32 embedded controller. Utilizing the classification model and only transmitting heartbeats classified as abnormal improves battery usage by 85.5% over streaming heartbeats to cloud services for analysis assuming 10% abnormal classifications per day at an average of 100,000 heartbeats per day.

Keywords: AI · IoMT · MIT-BIH · Deep Neural Network · Convolutional Neural Network

N. Filipović (Ed.): AAI 2023, LNNS 999, pp. 214–229, 2024.
https://doi.org/10.1007/978-3-031-60840-7_27

1 Introduction

1.1 The Problem

The increase in cardiac related injuries and deaths since the Covid-19 pandemic has been shown that there has been an increase of 23% to 34% of the excess death rates in those aged 25–44 years of age and 13% to 18% for those aged 45 and above [1]. There has also been an increase in myocarditis and pericarditis in young adults post Covid-19 [2]. According to the American Heart Association, heart disease and stroke kill more Americans than all forms of cancer and respiratory diseases combined [3]. New progress in machine learning and embedded technologies can provide a means to identify potential heart anomalies that would otherwise go unnoticed. New progress in machine learning and embedded technologies can facilitate long term health monitoring at home, hospitals, and assisted living facilities. Smart heart sensors provide a means to identify potential heart anomalies that would otherwise go unnoticed. The rapid acceptance of Internet of Things (IoT) and sensor technology provides opportunity for seamless integration into Internet of Medical Things (IoMT). New platforms, like smartwatches, integrated processing and communication platforms with ECG and PPG sensors. However, these signals are mostly utilized to provide basic measurements of heart rate or physical activity.

Previously, we developed one of the first real-time wearable monitors of myocardial ischemia [4] using DSP processor and analyzed power consumption and power optimization of the monitor [5]. However, state of technology did not allow prolonged monitoring. Recent developments of embedded AI and machine learning make it feasible to perform low-power, wireless monitoring in real-time [6].

Computationally heavy processing such as classifying and detecting abnormal heartbeats have not been integrated into these devices due to processor and power constraints. Instead, deeper analysis requires the data stream to be uploaded to a cloud-based server for an expert analysis of a cardiologist. A majority of the "wearables" today only provide limited feedback to the user through the watch interface or through a mobile application that records their beats per minute (BPM), inter-beat-interval (IBI), and in some cases even the heart rate variability (HRV). Integration of ECG into a smartwatch enabled detection of atrial fibrillation was proven to be possible, with a sensitivity of around 93% [7] using an Apple watch, but only when initiated by the user. However, it would be quite beneficial to be able to also process the data locally on the device to provide real-time analysis of any abnormal heart issues and then provide immediate alerts to the user while also uploading detailed measurements to a doctor for further analysis. One specific use case is an assisted living or nursing home facility in which patient health monitoring would be highly beneficial and would improve early detection of heart related issues.

1.2 IoMT and Wearable Issues

IoMT and wearable technologies must be small, highly energy efficient, have very limited resources in memory and computational capability, and are powered by a very small battery [8] These challenges pose major obstacles for moving machine learning to these devices. Machine learning models are typically computationally heavy and require large amounts of memory for storing the models and calculations which requires an increase in battery size, larger processors, and additional memory in the IoMT device. Today, data is continually being collected from IoMT devices and saved for later processing or uploaded to the cloud with minimal analysis on the device itself. It has been proven that the sensors themselves are highly reliable and accurate data sources for heart and health monitoring. WESAD, a Multimodal Dataset for Wearable Stress and Affect Detection [9] showed that simple wrist-based IoMT devices can provide a reliable data source of ECG signals. The additional development of Synergistic Personal Area Networks, like those proposed by Jovanov [10], combines multiple IoMT devices and sensors together to facilitate synergy of collected information. It has also been shown that these datasets can be utilized to create highly accurate machine learning models. As an example, Bangani et al. used machine-learning and artificial intelligence to develop an accurate stress detection model that utilize data recorded from IoMT devices to classify the stress level of nurses [11]. The next chapter in IoMT device evolution is the on-sensor or on-edge analysis and classification of measured data using machine learning on the devices themselves.

1.3 Advances in Embedded Machine Learning

There has been recent growth in moving machine learning to Internet of Things (IoT) and edge devices for use in speech recognition (Google Home, Amazon Alexa) and for facial recognition and augmented reality within smartphone applications. These enhancements have led to the development of specialized neural network coprocessors and machine learning models that are specially compiled for use on larger and more powerful embedded systems, like smartphones. Recent work [6, 12–14] in machine learning has created models that can identify and classify heart beat abnormalities from recorded data with high accuracy using Dense Neural Networks (DNN) and Convolutional Neural Networks (CNN). Training these models can take a few minutes to several hours depending on the size and complexity of the heartbeat ECG signals, the number of records, and because it is time-based, the window size of the data that is provided to the model for each sample. These challenges can easily be overcome even on a standard laptop.

However, it can still be a major challenge to deploy ML models on wearable and IoMT devices due to the lack of computational resources and energy usage. A new group of software engineers and corporations have been working to convert previously trained models into small and efficient software libraries that can be implemented on embedded systems called the TinyML Foundation [12]. TinyML is the study and development of machine learning on embedded systems specifically to overcome the energy, processing, and storage issues traditional models create for embedded systems. Since Python is the programming language of choice for machine learning development, the focus has been to modify the TensorFlow libraries into a "lite" version for IoT devices. Tensorflow Lite [15] is a software library created to specifically convert Python machine learning models for deployment on mobile, IoT, and edge devices. The goal is to be able to utilize the processing power of Python and TensorFlow, along with large datasets, to develop fully trained models on platforms with large processing power and memory. These trained models (weights) are then recompiled utilizing the Tensorflow-Lite libraries to reduce and optimize the models for smaller memory sizes. The models are then deployed onto microcontroller platforms and used for prediction only, since the training occurred previously on the larger platform.

2 Related Work

The following research articles were utilized in learning the state of the art in modeling of ECG data utilizing neural networks, PCA, and Attention-based networks. This information helped guide the direction for minimalizing and simplifying the types of neural networks that should be tested for an embedded system. Pramukantoro et al. tested multiple classification models and found Random Forest, decision-tree classifiers and artificial neural networks performed well using RR-intervals [16]. However, their device transmitted the data using Bluetooth Low Energy to a secondary platform for analysis. Wang et al. utilized Convolutional Neural Networks (CNN) and Long Short-term Memory (LSTM) models and found that neither model performed well with collected data and only performed around 90% accuracy [17]. A review of proposed models using CNN, LSTM and Support Vector Machines (SVM) classifiers for ECG signals found that performance and accuracy required extremely complicated models [18]. The size of these models would not fit in the memory of most embedded devices.

3 Proposed Solution

Recent advances in sensor and embedded system technologies and the advent of real-time data analytics and modeling allows the creation of devices to measure physiological changes and to accurately monitor heartbeats utilizing PPG and ECG signals in real time. Manufacturers have now developed systems-on-a-chip (SoC) that integrate ECG, PPG-Red and PPG-Infrared sensors on a tiny, embeddable sensor that will work with most microcontrollers [19]. These systems are non-invasive and can be integrated with wearable devices to provide real-time monitoring.

We propose the integration of an embeddable trained neural network that can identify heartbeat anomalies with IoMT wearables that is energy and computationally efficient. This requires that the model be small enough to fit within memory of the device and be able to process data in near real-time. Since the data is time-based, a buffer window must be identified that optimizes the accuracy of the model without being computationally or memory expensive. The ECG signal is stored into a buffer and the heartbeat peak is detected R-Peak detection using derivative-based with adaptive thresholding similar to Rodriguez et al. [20]. Modified derivative-based detectors that are optimized for embedded systems [21] were found to be computationally efficient [22]. Once the heartbeat peak has been identified, a new array of values equal to one second of measurements is generated with the peak in the center. This array is then sent to the classification model for processing. To simplify and reduce the size of the models, the first classifier will only classify a heartbeat as *normal* or *abnormal*. If a heartbeat is abnormal, the heartbeat data is sent to the secondary classification model to determine one of several abnormalities. In parallel the previous three heartbeats could be uploaded to the cloud for further analysis by a medical professional. This concept assumes that a majority of the time the heartbeat is normal and only when abnormalities are identified should further action occur, thus improving computational and energy usage (Fig. 1).

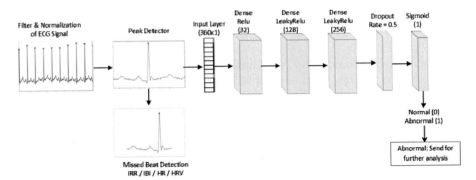

Fig. 1. Simplified DNN Architecture for heartbeat classification

3.1 Methods

The ECG is sampled as a continuous data stream at 360 Hz. The input features consist of all of the measured ECG values for a specific time duration window size. A derivative-based peak detector with adaptive threshold similar to Rodrigues et al. [21] was used to identify the peak on the input ECG signal to identify the heartbeat peak within the window of values. It was decided to evaluate each heartbeat "R-value", or peak, as the centralized data point with a fixed time window before and after the peak. The "before and after" window is then resized for time slices of 3 s, 2 s, 1 s, and 0.5 s for analysis to determine the highest accuracy for both model types. The input feature set is calculated by Eq. 1 below where f_s is the sample frequency and *window* is the window size in seconds. As an example, for a 3 s window (before and after) the feature set input to the model is 2,160 inputs as compared to 360 inputs for 0.5 s window (before and after). These different window sizes were chosen to determine the proper window size for optimal accuracy of the models (Figs. 2, 3, 4 and 5).

$$\text{Input\_dimension} = \text{window\_before} * \text{f\_s} + \text{window\_after} * \text{f\_s} \qquad (1)$$

Principal Component Analysis (PCA) was also evaluated as part of the testing of the optimized models to determine if the input to the models could be further reduced in comparison with the unaltered input stream. PCA was found to be computationally inefficient for little to no benefit in the model's accuracy.

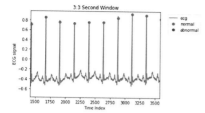

Fig. 2. 6 s window (2,160 features)

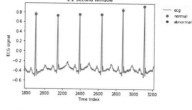

Fig. 3. 4 s window (1,440 features)

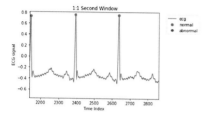

Fig. 4. 2 s window (720 features)

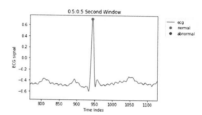

Fig. 5. 1 s window (360 features)

4 Dataset Overview

4.1 ECG Overview

Electrocardiogram is a measurement of the heart electrical activity as a graph of voltage versus time using electrodes placed on the skin. The electrodes measure the small electrical changes as the heart muscle contracts and relaxes during the cardiac cycle. The key parts of the ECG wave are the P-wave, the QRS complex and the T-wave. The P-wave represents the depolarization of the atria. The QRS complex shows the depolarization of the ventricles and the T-wave is the repolarization of the ventricles (Fig. 6).

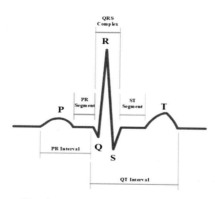

Fig. 6. ECG of a normal heart signal

4.2 MIT-BIH Dataset Description

The dataset used for this project is the MIT-BIH Arrythmia Database [23, 24]. The dataset contains 30-min recordings of a two-channel ambulatory ECG for 48 patients from 1979. The recordings were sampled at 360 Hz with 11-bit resolution. The datasets were annotated by two or more cardiologists with over 110,000 annotations that identify beats as normal, non-beat (22 types), or one of 18 abnormal types of heartbeats. Each dataset is broken out by patient with a continuous stream of 1-dimensional, time series data of ECG measurements (Fig. 7).

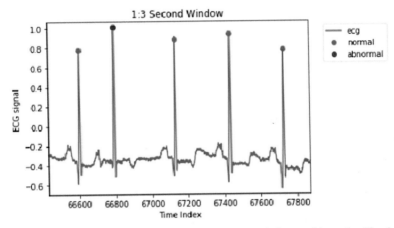

Fig. 7. MIT-BIH ECG sample data showing normal and abnormal beat classification

4.3 Data Preprocessing

The data is saved in individual patient files in the WaveForm DataBase (WFDB) [25] format as part of the MIT-BIH dataset which is included within PhysioBank. The WFDB format requires special software to read the data files and we utilized the wfdb-python [26] library to read and parse the data. The patient files are read and each heartbeat peak within the patient file is identified, along with its class. Based on the window size, a number of datapoints before and after the peak is combined to form a single data input array, with its corresponding class label. This list of points, with the heartbeat peak being at the center, becomes the input window size to the model. The first stage of the heartbeat analysis is to apply a filter to the ECG signal to reduce noise and then the data is normalized. Equation 2 is used to normalize the ECG signal.

$$ecg\_signal = (ecg\_signal - ecg\_signal.mean()) / ecg\_signal.std() \qquad (2)$$

The heartbeat peak detector utilizes the first derivative maximum to find the R-peak in the ECG signal to find the next R-Peak of the heartbeat. The sliding window is centered on the detected peak with a number of samples before and after the peak. Since the data is time-based, a buffer window is identified to optimize the accuracy of the model without being computationally or memory expensive. Next, missed beats are flagged based on Pan-Tompkins algorithm [20] that says the RR interval can only be in the range of 0.7 and 1.6 of the previous RR interval. For our model if the next beat interval is greater than 60% of the last interval, then a missed beat has been marked as detected. Since the data for each heartbeat is centered around the detected peak, the windowed data for each heartbeat, regardless of the separation between beats, is processed. If a missed beat is detected, an extended window of multiple beat times can be sent to the server for further analysis. The heartbeat data window is then sent to a Deep Neural Network (DNN) that identifies whether the beat is normal or abnormal. If the beat is normal, no further processing is performed. If the model classifies the heartbeat as abnormal, then the heartbeat data is passed to a second DNN to classify the type of abnormality. The previous 3 heartbeats can also be uploaded to the cloud for in-depth processing and analysis. This concept assumes that a majority of the time the heartbeat is normal and only when abnormalities are identified should further action occur. DNN models were selected for use in the embedded device over Convoluted Neural Networks due to their simpler design and reduced computational and memory requirements.

4.4 Classification Reduction

The dataset has 41 symbols to mark multiple types of heart-related issues. However, since the goal is to create a model that can be utilized on an embedded system, we are only interested in identifying an irregular heartbeat as a warning to the user that they should seek medical advice for the first level of classification. To reduce the dataset complexity, the classes were reduced to three: non-beat (3186), abnormal (34,409), and normal (75,052). The "non-beat" classification was dropped from the data. If the heartbeat is detected as abnormal, the second classifier groups the 41 symbols into six classes for identification of the heartbeat abnormality (Table 1).

Table 1. Class count by type

Count	Symbol	Meaning	Group as
75052	· *or* N	Normal beat	(Normal)
8075	L	Left bundle branch block beat	L
7259	R	Right bundle branch block beat	R
2546	A	Atrial premature beat	(Other)
150	a	Aberrated atrial premature beat	(Other)
0	J	Nodal (junctional) premature beat	(Other)
2	S	Supraventricular premature beat	(Other)
7130	V	Premature ventricular contraction	V
803	F	Fusion of ventricular and normal beat	(Other)
16	e	Atrial escape beat	(Other)
229	j	Nodal (junctional) escape beat	(Other)
106	E	Ventricular escape beat	(Other)
7028	/	Paced beat	/
982	f	Fusion of paced and normal beat	(Other)

5   Model Selection and Evaluation

5.1   Classification Models Selected

Three types of neural network classifiers were initially selected for testing with the datasets. Each of the models are intentionally designed to have only a few layers in order to keep their overall memory usage at a minimal because the model would be utilized in an embedded system after being trained. If the model has too many layers, the number of weights and the model complexity will increase the required memory needed to run the model. Deep Neural Networks (DNN), Convolutional Neural Networks (CNN), and Long Short-Term Memory networks were evaluated to determine the optimal model, epochs and window sizes for the models. After initial testing, it was decided that LSTM models were not a good fit for this dataset as the models never reached above 60% accuracy regardless of the window size of the feature set. For all models, the evaluation metrics would be based on classification accuracy using 30% of the dataset for testing.

5.2   Model Evaluation and Testing

To evaluate the best model and window sizes for the features, eight different window sizes were selected as seen in Table 2 below. The feature set window was to determine if having additional heartbeat peaks before or after the heartbeat being evaluated would be beneficial in increasing the accuracy of the models, and if so, how small can the window be to still be able to accurately predict abnormal heartbeats?

Table 2. Feature set window sizes

Feature Set	Seconds Before Peak	Seconds After Peak	Feature Input Size
3:3	3	3	2160
3:1	3	1	1440
2:2	2	2	1440
2:1	2	1	1080
1:1	1	1	720
1:2	1	2	1080
1:3	1	3	1440
0.5:0.5	0.5	0.5	360

For each window size, the DNN and CNN models were trained to identify the number of epochs needed to optimize each model. In each case, the models performed best between 8 to 10 epochs. For model evaluation it was decided to use 10 epochs for training. To evaluate the models and window sizes, the dataset was divided by patient into a training group and a test group with 70% of the patients being in the training group and 30% in the test group. It was decided to divide by patient instead of the dataset as a whole because an embedded heart monitor would be used on a patient-by-patient basis instead of a general set of data. If the models performed well after being trained on multiple patients and then being offered unseen patient data as the test values, the models should then perform well in a real-world application. It was found that the CNN model outperformed the DNN models in every category except when PCA was utilized to further reduce the feature size. It was also found that a single heartbeat profile (1 s window) was sufficient to accurately predict the heartbeat classification using the simplified CNN network with an accuracy of over 98.4%. We also determined that adding an Attention layer to the CNN model did not improve accuracy, but rather diminished it (Fig. 8).

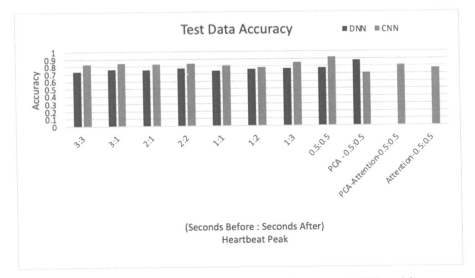

Fig. 8. Test data accuracy of each window size for DNN and CNN models

Focusing on the one second heartbeat window as model inputs and optimizing both the DNN and CNN models further, found that the DNN model could be optimized above 98% by normalizing the ECG signal and by adding two additional layers, where as the CNN model only performed around 96%. It was decided to move forward with the DNN model due to its reduced memory and computational size for the embedded devices over the CNN model. Preliminary testing of the CNN model on the embedded device found it to use four times more memory when compiled for embedded use. Figure 9 below describes the Deep Neural Network utilized for this research.

```
model = Sequential()
model.add(Dense(32, activation = 'relu', input_dim = X_train.shape[1]))
model.add(Dense(units = 128, activation=tf.keras.layers.LeakyReLU(alpha=0.001)))
model.add(Dense(units = 256, activation=tf.keras.layers.LeakyReLU(alpha=0.001)))
model.add(Dropout(rate = 0.5))
model.add(Dense(1, activation = 'sigmoid'))
model.compile(
        loss = 'binary_crossentropy',
        optimizer = 'adam',
        metrics = ['accuracy'])
```

Fig. 9. DNN: threshold = 0.32977

6 Results

Using the nursing home or assisted-living facility as the use-case, it was determined that the importance of catching all abnormal heartbeats was a higher priority over misclassifying a normal heartbeat as abnormal since any flagged heartbeats would be sent for further processing to a secondary model or to the cloud for human evaluation. Thus, the threshold for classification of normal/abnormal was modified from 0.33 to 0.01 (Figs. 10 and 11).

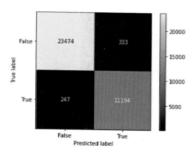
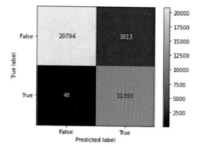

Fig. 10. DNN: threshold = 0.32977   **Fig. 11.** DNN: threshold = 0.01

This greatly reduced the likelihood of classifying abnormal beats as normal. This means that only about 0.136% of arrhythmias would be missed per day, but 99.9% of all arrhythmias would be detected and flagged for further analysis. Changing the threshold does increase the number of misclassed normal beats by a factor of 10, but the percentage of normal heartbeats misclassified as abnormal in a given day is around 8.6% of total heartbeats (assuming 100,000 heartbeats for a 24 h period) using the MIT-BIH patients' data as the comparison sample. These heartbeats would then be sent, along with properly classified arrhythmias, for further processing with the secondary model and the cloud.

The trained model was converted to Tensorflow-Lite [15] utilizing the *everywhereml* [27] Python library. This library converts the Python TensorFlow model to a C++ header file for use in embedded systems. The code was added to an Espressif ESP-32 microprocessor and tested utilizing multiple patient heartbeats from the test data to validate that the predict function of the model worked in the embedded system. The model accurately predicted each of the test heartbeats on the ESP-32 platform. To accurately determine the energy and time required for the model to predict the classification, the ESP-32 tested one of the test heartbeats every 5 s and the voltage, current, and power usage was measured using a JouleScope JS2200. It was found that it took 65.5 ms at 51 mw of power to classify a heartbeat as seen in Fig. 12 below. This confirms that the embedded algorithm can work in real-time and reduce the overall energy usage in comparison to sending every heartbeat over Wi-Fi for analysis.

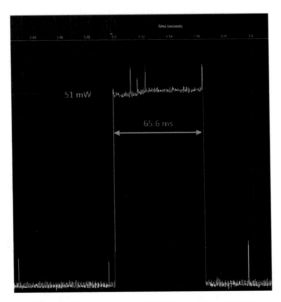

Fig. 12. Power usage of DNN model on ESP32

Further power analysis shows that transmitting each heartbeat of data over Wi-Fi using an Espressif ESP-32 using the MQTT [28] protocol utilizes an average of 22.27 milli-Coulombs of charge (6.19 μAh) as seen in Fig. 23. An average heart beats 100,000 times per day, so the required battery power to transmit the heartbeat data, at 360 Hz sampling rate for a 24-h period is 618.6 mAh. Our proposed solution classifies each heartbeat (0.276 μAh per heartbeat), but only requires beats classified as *abnormal* to be transmitted. The energy savings for each heartbeat classified as *normal* is 6.19 μAh – 0.276 μAh which is a savings of 5.91 μAh per heartbeat or 95.54%. Assuming that 10% of all heartbeats per day were classified as *abnormal*, the total energy used for heartbeat analysis over 24 h would drop from 618.6 mAh to 89.5 mAh for heartbeat analysis and transmission of *abnormal* classified heartbeats which is an energy savings of 85.54%. The lower number of false positives and lower number of detected arrhythmias in a day would improve the efficiency since they would not be required to be uploaded for further analysis (Fig. 13).

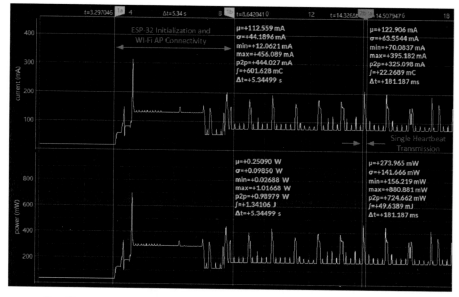

Fig. 13. ESP-32 Wi-Fi startup and heartbeat transmission power measurements

7   Conclusion

Heart disease is a leading cause of death throughout the world. In most cases, heart abnormalities are only identified after a major event occurs, or if a doctor happens to run a test during a routine visit. Having the ability to detect heartbeat abnormalities in real-time using simple IoT/IoMT wearables would identify many heart problems, stress disorders, and other health problems. In this research we have shown that Deep Neural Networks can be utilized on embedded devices to accurately identify missed beats and detect heartbeat abnormalities with accuracies over 99.9% with less than 0.136% of misclassified arrhythmias per day. Power savings of over 85% was realized based on an estimated 10% of classified *abnormal* beats per day. Future research opportunities focused on the "cost" in energy, time, and memory usage of the embedded models on the microcontrollers as well as live patient data analysis using the models should be performed.

References

1. Yeo, Y.H., et al.: Excess risk for acute myocardial infarction mortality during the COVID-19 pandemic. J. Med. Virol. **95**(1), e28187 (2023). https://doi.org/10.1002/jmv.28187
2. Akhtar, Z., Trent, M., Moa, A., Tan, T.C., Fröbert, O., MacIntyre, C.R.: The impact of COVID-19 and COVID vaccination on cardiovascular outcomes. Eur. Heart J. Suppl. **25**(Supplement_A), A42–A49 (2023). https://doi.org/10.1093/eurheartjsupp/suac123
3. Heart and Stroke Statistics: www.heart.org. https://www.heart.org/en/about-us/heart-and-str oke-association-statistics. Accessed 17 May 2023

4. Jovanov, E., Gelabert, P., Wheelock, B., Adhami, R., Smith, P.: Real time portable-heart monitoring using low power DSP. In: International Conference on Signal Processing Applications and Technology ICSPAT 2000, Dallas, October 2000
5. Jovanov, E., Raskovic, D., Martin, T., Hanief, S., Gelabert, P.: Energy profiling of DSP applications, a case study of an intelligent ECG monitor. In: International Conference on Signal Processing Applications and Technology ICSPAT 2000, Dallas, October 2000
6. Tsoukas, V., Boumpa, E., Giannakas, G., Kakarountas, A.: A review of machine learning and TinyML in healthcare. In: 25th Pan-Hellenic Conference on Informatics, Volos Greece, November 2021, pp. 69–73. ACM (2021). https://doi.org/10.1145/3503823.3503836
7. Pepplinkhuizen, S., et al.: Accuracy and clinical relevance of the single-lead Apple Watch electrocardiogram to identify atrial fibrillation. Cardiovasc. Digit. Health J. 3(6Suppl), S17–S22 (2022). https://doi.org/10.1016/j.cvdhj.2022.10.004
8. Martin, T., Jovanov, E., Raskovic, D.: Issues in wearable computing for medical monitoring applications: a case study of a wearable ECG monitoring device. In: Digest of Papers. Fourth International Symposium on Wearable Computers, October 2000, pp. 43–49 (2000). https://doi.org/10.1109/ISWC.2000.888463
9. Schmidt, P., Reiss, A., Duerichen, R., Marberger, C., Van Laerhoven, K.: Introducing WESAD, a multimodal dataset for wearable stress and affect detection. In: Proceedings of the 20th ACM International Conference on Multimodal Interaction, Boulder CO USA, October 2018. ACM, pp. 400–408 (2018). https://doi.org/10.1145/3242969.3242985
10. Jovanov, E.: Wearables meet IoT: synergistic personal area networks (SPANs). Sensors 19(19) (2019). https://doi.org/10.3390/s19194295. Art. no. 19
11. Bangani, R.G., Menon, V., Jovanov, E.: Personalized stress monitoring AI system for healthcare workers. In: 2021 IEEE International Conference on Bioinformatics and Biomedicine (BIBM), December 2021, pp. 2992–2997 (2021). https://doi.org/10.1109/BIBM52615.2021.9669321
12. TinyML Foundation. https://www.tinyml.org/. Accessed 20 Apr 2023
13. Alajlan, N.N., Ibrahim, D.M.: TinyML: enabling of inference deep learning models on ultra-low-power IoT edge devices for AI applications. Micromachines 13(6) (2022). https://doi.org/10.3390/mi13060851. Art. no. 6
14. Rahman, S., Khan, Y.A., Pratap Singh, Y., Ali, S.A., Wajid, M.: TinyML based classification of fetal heart rate using mother's abdominal ECG signal. In: 2022 5th International Conference on Multimedia, Signal Processing and Communication Technologies (IMPACT), November 2022, pp. 1–5 (2022). https://doi.org/10.1109/IMPACT55510.2022.10029140
15. TensorFlow Lite | ML for Mobile and Edge Devices. https://www.tensorflow.org/lite. Accessed 08 Apr 2023
16. Pramukantoro, E.S., Gofuku, A.: A heartbeat classifier for continuous prediction using a wearable device. Sensors 22(14) (2022). https://doi.org/10.3390/s22145080. Art. no. 14
17. Wang, P., et al.: A wearable ECG monitor for deep learning based real-time cardiovascular disease detection. arXiv, 24 January 2022. https://doi.org/10.48550/arXiv.2201.10083
18. Amin Ali, O.M., Wahhab Kareem, S., Mohammed, A.S.: Evaluation of electrocardiogram signals classification using CNN, SVM, and LSTM algorithm: a review. In: 2022 8th International Engineering Conference on Sustainable Technology and Development (IEC), February 2022, pp. 185–191 (2022). https://doi.org/10.1109/IEC54822.2022.9807511
19. MAX86150: Integrated photoplethysmogram and electrocardiogram bio-sensor module for mobile health. Mobile Health, p. 49
20. Pan, J., Tompkins, W.J.: A real-time QRS detection algorithm. IEEE Trans. Biomed. Eng. BME-32(3), 230–236 (1985). https://doi.org/10.1109/TBME.1985.325532
21. Rodrigues, T., Samoutphonh, S., Silva, H., Fred, A.: A low-complexity R-peak detection algorithm with adaptive thresholding for wearable devices. In: 2020 25th International Conference

on Pattern Recognition (ICPR), January 2021, pp. 1–8 (2021). https://doi.org/10.1109/ICP R48806.2021.9413245

22. Elgendi, M., Eskofier, B., Dokos, S., Abbott, D.: Revisiting QRS detection methodologies for portable, wearable, battery-operated, and wireless ECG systems. PLoS ONE **9**(1), e84018 (2014). https://doi.org/10.1371/journal.pone.0084018

23. Moody, G.B., Mark, R.G.: MIT-BIH arrhythmia database. physionet.org (1992). https://doi.org/10.13026/C2F305

24. PhysioNet Databases. https://physionet.org/about/database/. Accessed 09 Mar 2023

25. The WFDB Software Package. https://archive.physionet.org/physiotools/wfdb.shtml. Accessed 20 Apr 2023

26. The WFDB Python Package: MIT Laboratory for Computational Physiology, 20 April 2023. https://github.com/MIT-LCP/wfdb-python. Accessed 20 Apr 2023

27. eloquentarduino: EverywhereML, 15 April 2023. https://github.com/eloquentarduino/everyw hereml. Accessed 20 Apr 2023

28. test.mosquitto.org. https://test.mosquitto.org/. Accessed 28 Jan 2022

Is There a Legal Obligation to Use Artificial Intelligence in the European Union Accession Process?

Dragan Dakic[(✉)]

Faculty of Law, University of Kragujevac, 1 Jovana Cvijića Street, 34000 Kragujevac, Serbia
ddakic@jura.kg.ac.rs

Abstract. This research aims to examine whether there are any legal grounds to assert a responsibility on Serbia's part if the country does not utilize artificial intelligence (AI) in the process of joining the European Union (EU) even when there is no textual reference to such obligation. To comprehensively explore potential legal grounds for invoking this particular duty, we must first establish the applicable legal frameworks in Serbia. We shall investigate the formal grounds of Serbia – European Union association process, which were established by the Stabilisation and Association Agreement, in the first part of the study. The focal point of the inquiry revolves around the significance attributed to this agreement within the Serbian domestic legislation. The objective is to determine if Serbia can potentially avoid its international obligation to employ AI based on its internal laws. This will help us in identifying the nature of the obligations Serbia may possibly have in this context, as well as the general principles guiding their interpretation. Subsequently, we will analyse the specific obligations that Serbia, as a candidate country, currently possesses during the integration stage. In the next section of the paper, we will highlight the inadequate fulfilment of some of the main requirements of the process, in order to better understand whether AI could assist in meeting these obligations and what characteristics such AI would have. This will lead us to the central research question at hand – whether there is a duty to utilize AI. We examined the rules pertaining to the possibility of evasion due to impediments, in contrast to the principle of good faith performance, in order to draw our conclusions.

Keywords: legal obligation · use · AI · EU accession

1 Introduction

The EU accession is an intricate and multi-step course of action encompassing political, economic, and legal aspects. An indication of the politically oriented nature of the accession process, the legal basis for EU accession is a single treaty article that offers limited direction. As the accession process has evolved, there has been a significant emphasis on conditionality in the EU's enlargement strategy, along with the introduction of political and economic criteria known as the 'Copenhagen criteria'. The criteria for membership

© The Author(s), under exclusive license to Springer Nature Switzerland AG 2024
N. Filipović (Ed.): AAI 2023, LNNS 999, pp. 230–247, 2024.
https://doi.org/10.1007/978-3-031-60840-7_28

in the European Union, known as the Copenhagen criteria, were established at the 1993 Copenhagen Summit. These criteria encompass political, economic, and administrative aspects. Specifically, a country seeking EU membership must demonstrate:[1]

1) the establishment of stable institutions ensuring democracy, the rule of law, the protection of human rights, and minority rights;
2) the development of a market economy capable of withstanding competition and market pressures within the EU; and
3) the capability to fulfil the responsibilities of membership, including the commitment to the objectives of political, economic, and monetary union.

Article 49 of the Treaty on the EU serves as the foundation for the accession process, outlining the eligibility requirements for applying for the EU membership and the procedure for becoming a member. The European Council's eligibility criteria must also be taken into account. The candidate country must demonstrate respect for and dedication to promoting the Union's values as outlined in Article 2 of TEU.[2] The accession process entails the adoption of the existing EU law, preparations to effectively apply and enforce it, and the implementation of various reforms in areas such as judiciary, administration, economy, and more. These reforms are necessary for the country to meet the conditions and criteria for joining, commonly referred to as the accession criteria. The accession process is divided into distinct stages, including pre-accession, accession negotiations, and post-accession monitoring. Throughout the process, applicants are subject to increased scrutiny. The political criteria for accession involve the stability of institutions that ensure democratic principles, the rule of law, protection of human rights, and the safeguarding of minorities. The economic criteria encompass the functioning of a market economy and the ability to handle competition and market forces. Additionally, the candidate country must exhibit the ability to fulfil the obligations of membership and enforce the EU *acquis*.

The general overview given above indicates that political, economic, and legal aspects of the process are not clearly distinguishable. Furthermore, if either of those aspects would be viewed in isolation from the remaining two, it would still maintain its heterogeneous nature. In this regard, a closer look at the legal aspect of the accession process reviles its complex structure. It introduces obligations to the candidates that could be differently categorized depending on criteria applied. In our view, the most purposeful categorization of the concerning obligations could be based on their effects. Following this criterion, the legal aspect imposes obligations that could be noted as declarative - referring to those like dedication to promoting the Union's values. Those are mostly indirect and descriptive obligations of the means. Another sort of obligation could be noted as absolute - referring to those such as the adoption of the existing EU law, its enforcement, and fulfilment of the EU obligations, etc. Absolute obligations are certainly more strict and measurable. As such they should be understood as obligations of the result. This categorization could be applicable to the candidate's obligations

---

[1] Accession criteria (Copenhagen criteria), https://eur-lex.europa.eu/EN/legal-content/glossary/accession-criteria-copenhagen-criteria.html.

[2] Consolidated version of the Treaty on European Union https://www.legislation.gov.uk/eut/teu/article/2.

concerning the political, economic, and legal aspects of the accession process. Namely, all three aspects of the process are covered with appropriate agreements between the European Communities and their Member States on the one part, and the candidate, on the other part.

The idea of this particular research is to investigate if any of those aspects are providing legal grounds to claim a duty on the part of Serbia to use AI in the EU accession process. In order to comprehensively research potential legal grounds to invoke this specific duty, first we need to define legal frameworks applicable to Serbia. In this part of the research we will investigate the formal grounds of the Serbia – European Union association process which were framed by the Stabilisation and Association Agreement. The central part of the inquiry concerns the status conferred to this agreement within Serbian internal law. The intention is to see if Serbia can evade its eventual international obligation to use AI based on its internal law. This should further help us to identify the type of obligations that Serbia eventually has in this regard and general rules in their interpretation. Following this part, we are going to analyse particular obligations that Serbia as a candidate country has at the current stage of the integration. In this part of the paper, we will point to the underperformance concerning some of the main requirements of the process as we could better understand if AI could help to meet those obligations and what would be the features of purposed AI. This should introduce us into the discussion about the central research question herein - is there an obligation to use AI? Rules applicable to the evasion due to an impediment in contrast to the principle of good faith performance were cross-examined to reach the conclusion.

2   Serbia and the European Union

Integration of the Western Balkans within the EU is framed within a special legal regime known as the Stabilisation and Association Process (SAP) which is followed by Stabilisation and Association Agreements.[3] Those are bilateral, far-reaching agreements that introduce a contractual relationship with the EU, entailing mutual rights and obligations. The Stabilisation and Association Agreement between the European Communities and their Member States on the one part, and the Republic of Serbia (SAA), on the other part was signed in 2008,[4] and it officially came into effect in 2013 after it was finally ratified at all instances within the EU (Lithuania was the last to do so). The main objective of this agreement is to establish a strong and enduring partnership between the EU and Serbia, based on the principles of reciprocity and mutual interest. A wide range of policy areas are covered by this agreement, including economic and trade policies, industrial cooperation, education, and training, all aimed at promoting Serbia's development and growth potential.

As to the SAA from the EU law perspective, it is an agreement specifically tailored for the countries of the Western Balkans and it represents the third generation of EU

---

[3] Stabilisation and Association Process - European Commission (europa.eu).

[4] ЗакОн о потврђивању Споразума о стабилизацији и придруживању између европских заједница и њихових држава чланица, са једне стране, и Републике Србије, са друге стране (Службени гласник РС – Међународни уговори, бр. 83/2008).

association agreements.[5] In line with other agreements from this group, SAA is characterized as preferential trade agreements[6] that also include political clauses.[7] Like it was previously noted, primary purpose of the SAA was to enable Serbia to become a candidate for EU membership; however, it should be noted that admission to the EU is not an automatic right. Within the hierarchy of the EU legal system, Stabilisation and Association Agreements are generally categorized as primary sources, along with other international treaties. They hold a higher legal force compared to secondary legislation, solidifying their crucial role within EU Law. Consequently, no transposition measures are necessary for their implementation.[8] The direct effect of Stabilisation and Association Agreements provisions on the establishment of free trade zones, which promote the smooth movement of capital, goods, and services between the associated states and the EU and its Member States, has been acknowledged by the EU Court of Justice. Professor Radivojević considers that suchlike recognition has practical implications, providing citizens and legal entities from Serbia with certain rights within the EU legal system.[9] Consequently, individuals and organizations will have the ability to directly invoke the provisions of the SAA during legal proceedings before the courts of the Member States and the EU Court of Justice, in line with standards similar to those enjoyed by EU citizens.

For the purpose of our inquiry, it is important to understand the status of the SAA within the Serbian legal system. Since it is a Continental Law country, the status of the international treaties was decided by the Serbian Constitution. In the relevant part, the Constitution stipulates (Article 16 Sect. 2) that: "Generally accepted rules of international law and ratified international treaties shall be an integral part of the legal system in the Republic of Serbia and applied directly."[10] At this point, it appears to be clear that SAA as an international ratified treaty, is perceived as an integral part of the domestic legal system with direct effects. Suchlike a position would be comparable to that within the EU. However, other Constitutional provisions are also relevant for SAA status determination. In the very same Article 16 Sect. 3, the Constitution introduces the requirement that ratified international treaties must be in accordance with the Constitution. Also in

---

[5] Радивојевић, З., & Radivojević, Z. (2012). Споразуми о стабилизацији и придруживању у праву Европске уније. *Zbornik radova Pravnog fakulteta u Nišu*, 62, стр. 53–72. http://www.prafak.ni.ac.rs/files/zbornik/sadrzaj/zbornici/z62/04z62.pdf.

[6] See Article 9 of the SAA CELEX:22013A1018(01):EN:TXT.pdf (europa.eu).

[7] See Title II of the SAA CELEX:22013A1018(01):EN:TXT.pdf (europa.eu).

[8] 12/86, Meryem Demirel v. Stadt Schwabisch Gmünd (ECR, 1987, p.3719), 192/89, Sevince v. Staatssecretaris van Justicie, (ECR I, 1990, p.3641); 237/91, Kus (ECR I, 1992, p.6781); 432/92, The Qeen v. Minister of Agriculture, Fisheries and Food (ECR I, 1994, p.3087) и 355/93, Eroglu v. Land Baden Wurrtenberg (ECR I, 1994, p.5113).

[9] Радивојевић, З. и Radivojević, Z., 2012. Споразуми о стабилизацији и придруживању у праву Европске уније. *Zbornik radova Pravnog fakulteta u Nišu*, [на Интернету] (62), стр.стр. 53–72. Available at: < http://www.prafak.ni.ac.rs/files/zbornik/sadrzaj/zbornici/z62/04z62.pdf >.

[10] Constitution of Republic of Serbia, available at http://www.parlament.gov.rs/upload/documents/Constitution_%20of_Serbia_pdf.pdf.

its section (sub)titled "Hierarchy of domestic and international general legal acts" the Constitution in its Article 194 Subsect. (4) introduces that:[11]

"Ratified international treaties and generally accepted rules of the international law shall be part of the legal system of the Republic of Serbia. Ratified international treaties may not be in noncompliance with the Constitution."

This led some authors to the conclusion that the SAA was subrogated against national Constitution.[12] Suchlike status determination could be the sort of reduced national-law framed reasoning. Accordingly, the SAA's status is mostly decided on the grounds of its hierarchal position in domestic legal system. In that sense, the SAA is perceived to be higher ranked legal act when compared to domestic laws. One of the arguments in support of this view could be relied upon the fact that national laws were subjected to the process of conformation with the SAA. Also, supremacy of the SAA over domestic laws is arising from the twofold effects of its direct application. First, parties can bring their cases before national civil and administrative courts due to the violation of their rights covered by the SAA. Second, national courts are obliged to apply provisions of SAA even if those provisions are in collision with domestic laws. Comparable reasoning was taken by the European Court of Justice in respect to the conflict between the EU law and national law.[13]

However, discussion about the status conferred to the SAA within the national legal system cannot be sustainably viewed only from the prism of the national law. Rather, its status should be interpreted in the broader context of the dualistic–monistic controversy as well as in the context of the relevant rules of international law and existing practice. In the most general terms, dualism strictly distinguishes between national (municipal) and international law. Those two distinctive legal systems have different formal and substantive sources. As to the formal sources, the sources of International Law are international legislative treaties[14] while the sources of Municipal Law are national statutes. As to the substantive sources, International Law is sourced out of the consensus between at least several states, while Municipal Law is introduced through the internal legislative procedure with preclusive competencies of the national bodies. Additionally, addressees of the International Law are states whereas the domestic aspects of government and matters between individuals, as well as between individuals and the administrative apparatus, are governed by Municipal Law.[15] As we can see, dualism leaves no space for debate about international law status within the national legal system since internal issues are out of

---

[11] Ibid.

[12] Vukadinović, R.D., Вукадиновић, Р.Д., Vukadinović Marković, J. и Вукадиновић Марковић, J., 2020. Uvod u institucije i pravo Evropske unije. Beograd: Službeni glasnik.стр.402.

[13] Judgment of the Court of 9 March 1978. Amministrazione delle Finanze dello Stato v Simmenthal SpA. Reference for a preliminary ruling: Pretura di Susa - Italy. Discarding by the national court of a law contrary to Community law. Case 106/77.

[14] For more details refer to Article 38 of the Statute of the International Court of Justice. Available at https://www.icj-cij.org/index.php/statute.

[15] Shaw, M. N. (2021) International Law. 9th edn. Cambridge: Cambridge University Press. https://doi.org/10.1017/9781108774802.

the scope of its domain. In order to operate within the domestic legal system dualistic approach requires express legislative incorporation of an international agreement.

Unlike that, the essential presumption of the monistic approach is substantive unity between national and international law. According to this approach, International Law and internal law are merely two branches of the unique legal system. Legislative incorporation is not a requirement for the operability of international agreements within the domestic legal system. However, scholarship of the monistic approach is concerned with the hierarchy between International Law and internal law.[16] It might be so because this aspect of their relationship is crucially important for the appropriate assessment of state compliance with its international obligations. In that sense, the supremacy of International Law could be considered a dominantly accepted view within a monistic approach.[17] Rationale for suchlike a theoretical position could be firmly supported by the practice of the International Court of Justice (ICJ) i.e. its predecessor the Permanent Court of International Justice (PICJ). Namely, the contractual parties are forbidden by the PCIJ from using internal regulations as a means to evade international obligations.[18] In terms of the hierarchy between domestic and international law, the PCIJ has established that constitutional provisions also fall under the mentioned restriction.[19] Regarding the interpretation of international agreements, the primary focus is on the intention of the contracting parties. This stance is clearly expressed by the PCIJ in its ruling on the Jurisdiction of the Courts of Danzig.[20] Herein the PCIJ stated that it is beyond doubt that the objective of an international agreement, as intended by the contracting parties, is to establish definitive rules that confer individual rights and obligations, and which are enforceable by national courts.[21]

All of this perfectly describes the situation with SAA in many aspects. Additionally, Serbia is regarded as a country that embraces a monistic approach.[22] Unavoidably, this should be considered when addressing the SAA and its status. From the perspective of this occurrence, the Constitutional provision consisted in Article 16 Sect. (3) in conjunction with the Article 194 Sect. (4) and Article 167 Sect. (2) which essentially requires that ratified international treaties may not be in noncompliance with the Constitution,[23] could be differently interpreted. As we have seen, ratified international treaties and generally accepted rules of international law are considered to be a part of the Serbian legal system. Also, in Article 142 the Constitution obliges national courts to apply enumerated

---

[16] Крећа, М. и Креħа, М., 2023. *Међународно јавно право*. 14. допуњено изд. изд. Београд: Правни факултет Универзитета, Центар за издаваштво.стр.844 стр.

[17] *Ibid.*

[18] Permanent Court of International Justice, Ser. B., No. 17, pp 32–33.

[19] Permanent Court of International Justice, Ser. A/B, No. 44, pp 24.

[20] Permanent Court of International Justice, Jurisdiction of the Courts of Danzig, Ser. B, 15, 1928, p. 18.

[21] *Ibid.*

[22] Милисавље Евић, Б., Палевић, М. и Palević, M., 2017. Однос међународног и унутрашњег права према Уставу Републике Србије из 2006. године. *Српска политичка мисао*, [на Интернету] (издање), стр.Стр. 29–46. Available at: < http://www.spmbeograd.rs/arhiva/SPM-2017-posebno/CD%20SPM%20SPC%202017.pdf >.

[23] Constitution of Republic of Serbia, available at http://www.parlament.gov.rs/upload/documents/Constitution_%20of_Serbia_pdf.pdf.

international law sources.[24] No doubt, the SAA has direct effects in Serbia and in the event of its collision with domestic law it shall have priority. Still, concerning requirement for Constitutional compliance with international treaties remains. Accordingly, it truly appears as if the Constitution is highly ranked as compared to the SAA. However this apparent hierarchical disposition of the SAA is rather declarative and has no practical nor even legal implications for the following reasons.

In the temporal sense, eventual assessment of the SAA compliance with the Constitution is procedurally possible only in the post-ratification period. It is important to note that the process of international treaty ratification entails various sequential procedures in Serbia. Following the signing of an international treaty, the National Assembly of Serbia must give its consent through ratification. The initiation of the ratification process is attributed to the Ministry of Foreign Affairs, responsible for submitting the treaty to the National Assembly for deliberation and endorsement. So, at this point, the State has undoubtedly expressed its consent to be bound by international contract which makes it almost impossible to invoke provisions of internal law to invalidate it. Namely, according to Articles 46 and 47 of the 1969 Vienna Convention, it is possible to invalidate international treaties based on the provisions of internal law. But this possibility is exceptional and it only refers to the provisions of internal law regulating procedural competencies: regarding competence to conclude treaties (Article 46), and specific restrictions on authority to express the consent of a State (Article 47). Either way, substantive non-compliance to the national Constitution, even if diagnosed by the Constitutional court, is not accepted ground for international treaty invalidation under the provisions of the 1969 Vienna Convention.

As a matter of general rule, a party may not invoke the provisions of its internal law, including the Constitution, as justification for its failure to perform a treaty (Article 27).[25] This was clarified in PICJ case law as we have previously seen. The general rule concerned was also reflected in the Articles of Responsibility of States for Internationally Wrongful Acts from 2001.[26] Finally, the legal irrelevance of the Constitutional provisions was formally verified in the Law on the Constitutional Court.[27] In the relevant part, this Law recognizes the higher relevance of international law over constitutionality. For instance, in Article 58 the Law introduces that particular clauses of an international treaty which were found to be incompliant to the Constitution are casing to be effective but only if that scenario is possible under the provisions of the treaty itself and general rules of international law. Therefore, substantive non-constitutionality of international

---

[24] *Ibid.*

[25] Vienna Convention on the Law of Treaties 1969, available at https://legal.un.org/ilc/texts/instruments/english/conventions/1_1_1969.pdf.

[26] Text adopted by the Commission at its fifty-third session, in 2001, and submitted to the General Assembly as a part of the Commission's report covering the work of that session. The report, which also contains commentaries on the draft articles, appears in <u>Yearbook</u> of the International Law Commission, 2001, vol. II (Part Two). Text reproduced as it appears in the annex to General Assembly resolution 56/83 of 12 December 2001, and corrected by document A/56/49(Vol. I)/Corr.4 Available at https://legal.un.org/ilc/texts/instruments/english/draft_articles/9_6_2001.pdf.

[27] (Official Gazette of the Republic of Serbia, No. 109/2007, 99/2011, 18/2013 - CC decision, 103/2015, 40/2015 - other laws, 10/2023 and 92/2023).

treaties in itself cannot deprive international treaties of their binding force. This firmly reaffirms the principle of relative irrelevance of internal law in the area of international treaties. Provisions of this Law should not be isolated from the context of the Opinion on the Constitution of Serbia of the Venice Commission[28] which took a critical stance in respect to observed Constitutional provisions.

3 Accession Negotiation Obligations

Serbia obtained the status of an EU candidate in March 2012 and has since opened 18 negotiation chapters, with 2 chapters already provisionally closed. Currently, the accession process is sequenced into 35 clusters with 22 opened to negotiation. This phase is marked with the generic notion of Accession Negotiation. At this stage, Serbian legal obligations in respect to the EU are arising from SAA, the Opening Statement of the Republic of Serbia,[29] Action Plans,[30] etc. In fact, the whole Accession Negotiation should be understood as an agreement on the terms, the necessary measures and ways of Serbian compliance to its obligations.[31] As we could properly decide which sorts of obligations (declarative/of the conduct or absolute/of the result) are dominant at this phase, we should recall that the Copenhagen criterion requires that a country seeking EU membership must demonstrate its political, economic, and administrative readiness without focusing on concrete results. Based on its previous experiences in the enlargement process, the EU created different negotiation frameworks for the Western Balkan countries. According to this new approach, a country seeking EU membership must prove concrete results achieved in the areas that are a priority for the negotiations. In the case of Serbia, priority areas are the rule of law, reform of public administration and local self-government and economic management.[32] According to Professor Miščević, this is 'the first time a candidate country must prove, before the accession, all three phases of harmonization of national law with EU law - harmonization, implementation and enforcement.'[33] The requirement for concrete results is also evident in the wording of Article 72 of SAA about the approximation of the existing legislation in Serbia which introduces that.

> (…) Serbia shall ensure that existing and future legislation will be properly implemented and enforced.[34]

---

[28] European Commission for Democracy through Law (Venice Commission), Opinion on the Constitution of Serbia, No. 405/2006, CDL – AD, 2007 para 14 – 18. Available at https://www.venice.coe.int/webforms/documents/default.aspx?pdffile=CDL-AD(2007)004-e.

[29] The Opening Statement of the Republic of Serbia, Intergovernmental Conference on the Accession of the Republic of Serbia to the EU, 21 January 2014.

[30] Action Plans.

[31] Miščević, T. и Мишчевић, T., 2018. Legislative obligations of Serbia in the accession process to the European Union - the case of fundamental rights. Yearbook. Provincial Protector of Citizens - Ombudsman; Institute of Criminological and Sociological Research.стр.333–348.

[32] Ibid.

[33] Ibid.

[34] Article 72 of the SAA CELEX:22013A1018(01):EN:TXT.pdf (europa.eu).

Therefore, it appears that the most dominant proportion of Serbian obligations at the Accession Negotiation stage are the obligations of the result in their nature. Also, monitoring mechanisms applied in this stage are designed to track records of concrete results.[35] If we accept this conclusion, we should further see in which way AI could help Serbia in meeting its priority areas obligations. This will introduce us to the possible answer to the question if absolute obligations can bring a duty to use AI.

4   Can AI Help?

It's been more than 10 years since Serbia obtained the status of an EU candidate and it still stands distant from the membership. Presumably, some additional help and innovative approaches are needed in this regard. If we ask Open.ai the question: 'Can we use Artificial Intelligence to accelerate Serbia's European Union Accession?' its answer would identify AI as a potentially supportive tool due to its perceived capabilities to process and analyse large amounts of data, identify necessary reforms, optimize processes, and enhance transparency. Simultaneously, ChatGPT-4 identified AI's efficiency as largely dependent upon data availability and quality, algorithm accuracy, and stakeholder willingness to adopt and use the technology. All of this indicates that AI could be a helpful optimizing tool in many aspects of Accession Negotiation. This could be well demonstrated on the example of the judiciary - one segment of the priority areas. Reasons to discuss potential AI application in the judiciary are threefold.

First, the National Strategy for the Development of Artificial Intelligence (2020–2025) sets forth measures for improving public sector services using artificial intelligence (Measure 4.5). Second, Cluster 1 of the Negotiation Process posed specific requirements concerning Judiciary and Fundamental Rights (former Chapter 23) and Justice, Freedom, and Security (former Chapter 24) firmly relying on human rights standards introduced through the European Convention on Human Rights and Fundamental Freedoms (EcHR).[36] Third, the EcHR incompliant judicial/administrative decisions are among the root causes of a large number of domestic cases pending before the European Court of Human Rights (ECtHR). This is considered to be a significant obstacle to the EU accession process as a whole, since at this stage of the Accession Negotiation fundamental rights evolved into a fundamental, pivotal, and indispensable component of the membership negotiation process incorporated in the Rule of Law chapters/clusters.[37]

Building and implementing supportive and suggestive AI purposed to provide case-customized information about human rights requirements i.e. that enables judges/public officials to deliver EcHR-compliant decisions, would increase the efficiency of the judiciary and domestic remedies, and lower the time and the cost of proceedings. No doubt such configured AI would be an accelerator of Accession Negotiation enabling Serbia to meet its obligations.

---

[35] See Serbia report 2023. Available at https://neighbourhood-enlargement.ec.europa.eu/serbia-report-2023_en.

[36] The European Convention on Human Rights and Fundamental Freedoms, https://www.echr.coe.int/documents/d/echr/convention_eng.

[37] See Miščević, T. и Мишчевић, Т., 2018. Legislative obligations of Serbia in the accession process to the European Union - the case of fundamental rights. Yearbook. Provincial Protector of Citizens - Ombudsman; Institute of Criminological and Sociological Research.стр.333–348.

5   What Are the Software Features and Purpose?

With recent progress in the development of AI, the technical aspect appears less problematic even though software's capabilities could be reliably confirmed or dismissed only through empirical methods that imply its actual implementation in the judicial system and its effective application to a sufficient extent in decision-making. In the technical sense, the most challenging task would be to overcome the disciplinary gap between legal norms and algorithmic codes i.e. to investigate and allocate the most appropriate module(s) for transforming legal norms into AI implantation data. Due to the vital relevance of data produced, obtained, collected, and generated through legal analysis (norms, standards, case law), adjusted techniques of legal data transformation into machine readable codes, such as those used in spectral clustering algorithms, need to be developed based on the experiences of the development/application of AI models in medicine including but not limited to neural networks, (hybrid) medical records for AI and their role in diagnostics, etc.

Prior to this phase, however, some theoretical and, perhaps, legislative preparatory work might be necessary. The starting point herein would essentially be about assessing the responsiveness of the current legal and institutional landscape toward AI introduction into the public sector (administration and judiciary). Even though institutional responsiveness can be tenably presumed since Serbia made significant progress in the field of digitalization and structural adjustments to it, this aspect should not be entirely neglected. As to the legal responsiveness, deeper analysis of internal legal requirements would be mandatory. The inclusion of the optics of the EU legal standards operating in the priority areas identified in the enlargement negotiations would be mandatory. Since detailed legal analysis would out-scope this paper, we are just going to provide a list of the accounts necessary for the comprehensive assessment:

- determination of normative and institutional responsiveness for the AI application in judicial decisioning in the field of civil law, the cases concerning family affairs focusing on anti-discrimination and protection of the rights of vulnerable groups (as they were defined through the documents of the Council of Europe);
- determination of normative and institutional responsiveness for the AI application in judicial decisioning in the field of civil law, the cases concerning status rights focusing on anti-discrimination and protection of the rights of vulnerable groups;
- determination of normative and institutional responsiveness for the AI application in judicial decisioning in the field of civil law, the cases concerning property rights focusing on anti-discrimination and protection of the rights of vulnerable groups;
- determination of normative and institutional responsiveness for the AI application in judicial decisioning in the field of administrative law, the cases concerning property rights focusing on anti-discrimination and protection of the rights of vulnerable groups;
- determination of normative and institutional responsiveness for the AI application in judicial decisioning in the field of civil or administrative law, the cases concerning freedom of expression and freedom of the media;
- determination of normative and institutional responsiveness for the AI application in judicial decisioning in the field of criminal law, the cases concerning fight against corruption and economic crime;

- determination of normative and institutional responsiveness for the AI application in judicial decisioning in the field of criminal law, the cases concerning vulnerable groups.

Along with the abovementioned responsiveness assessment, comprehensive legal analysis necessarily needs to encompass an assessment of technical faculties that the proposed AI should have in order to be suitable for official use. This conformity aspect would also be a combination analysis of national statutory requirements concerning technical faculties of AI together with an assessment of its alignment to human rights requirements applicable in listed proceedings.[38]

Potential challenges to the proposed interdisciplinary investigation of AI's capabilities to accelerate EU accession are associated with its novelty both in theoretical and applicative aspects. Theoretical debates about AI applications in decision-making are dominantly centred on general legal issues such as civil liability, tort, consent, and privacy.[39] The vast majority of publications are devoted to AI in medicine while research on AI in judiciary, beyond the context of the rule of law, is largely missing. It is so even regarding general issues including normative and institutional landscape for AI introduction. As a result, doctrinal frameworks concerning AI application in the judiciary are poor. As to the applicative aspect, up to date, people managed to convert natural voice into codes (in real-time) or medical parameters into AI implantation data, but technology as described herein appears to be missing.

6 Is There an Obligation to Use AI?

As we have seen Serbian obligations from the Accession Negotiation are that of the result. In order to successfully meet them and align with the EU *acquis,* Serbia must achieve EcHR standards in its judicial and administrative practice. In the terms of measurability, this root cause of a large number of domestic cases pending before the ECtH, as one of the main obstacles to the EU accession process as a whole, must be eliminated. Two additional circumstances emphasize the character of Serbian obligations and position. On the one hand, Serbia cannot invoke internal law to invalidate, postpone, or avoid the concerning obligation. On the other hand, purposed AI could enable judges/public officials to deliver EcHR-compliant decisions, and help Serbia meet this obligation.

---

[38] Dakić, D. (2023). Reproductive Autonomy Conformity Assessment of Purposed AI System. In: Filipovic, N. (eds) Applied Artificial Intelligence: Medicine, Biology, Chemistry, Financial, Games, Engineering. AAI 2022. Lecture Notes in Networks and Systems, vol 659. Springer, Cham. https://doi.org/10.1007/978-3-031-29717-5_3.

[39] ДАкић, Д. и Dakić, D., 2023. Аватари као пружаоци услуга: међународноправни аспекти. [на Интернету] Правна регулатива услуга у националним законодавствима и праву Европске Уније: [зборник реферата по позиву са Међународног научног скупа одржаног 19. и 20. маја 2023. године, на Правном факултету у Крагујевцу у организацији Института за правне и друштвене науке Правног факултета Универзитета у Крагујевцу]. Правни факултет Универзитета, Институт за правне и друштвене науке.стр.‡стр. ‡399–412. Available at: < http://institut.jura.kg.ac.rs/images/Projekti/7%20XXI%20vek/Pravna%20regulativa%20usluga%20u%20nacionalnim%20zakonodavstvima%20i%20pravu%20EU%202023.pdf.

At this point, it appears that Serbia undoubtedly has an absolute obligation to use AI as described here. Adding to this *bona fides* principle that mandates treaty parties to fulfil their obligations in good faith and abstain from actions that undermine the intended goals and objectives of the treaty,[40] Serbian obligation to use AI seems inevitable. But before the conclusion, we should consider how potential hardships could affect presumed obligation. In the beginning, it is tenable to preclude from the list of excusing impediments such as lack of doctrinal and normative foundations, or eventual institutional non-responsiveness. It is so because all of them are attributable to one of the contracting parties. As such, those shortcomings cannot afford the party affected ground for invoking the termination or suspension of the treaty nor obligations arising from it.[41]

The situation, however, might differ if purposed AI itself is missing. In that event, impossibility of performance that essentially serves as a means to provide relief to a promisor when their contractual performance deviates significantly from what was reasonably anticipated of them might arise. This relieving doctrine has been theoretically discussed and normatively reflected in the 1969 Vienna Convention as well as in some so-called irregular sources of law.

In brief, theoretical considerations are centred on the conditions that trigger the application of this doctrine. It is understandable since any departure from the rule *pacta sunt servanda* needs to be restrictively interpreted and well justified. Among the excusing factors, scholarship provided an objective approach that argues that the doctrine should only apply when performance becomes objectively impossible. In contrast to the former, a subjective approach suggests that impracticability should be judged from the perspective of the party unable to perform. Perhaps those factors are not necessarily mutually excluding since they are usually co-joint in practice. Theory recognized this when noted that "[a] thing is impossible in legal contemplation when it is not practicable, and a thing is impracticable when it can only be done at an excessive and unreasonable cost."[42] Cost-qualifying notions 'excessive' and 'unreasonable' could be both subjectively and objectively interpreted. Either way, if we take the cost as the criterion of the impossibility of performance, the situation as follows. Faculty of Technical Sciences Čačak of the University of Kragujevac had offered to develop the software capable to:[43]

- overcome disciplinary gap between legal norms and algorithmic codes i.e. to investigate and allocate the most appropriate module(s) for transforming legal norms into AI implantation data that are going to be tested/used/upgraded through the software development relying on mathematics and AI science;
- provide case-customized information about human rights requirements to the judges/public officials enabling them to deliver ECHR compliant decision;
- provide open access to the software as all interested persons could learn ECHR guarantees applicable to particular case/circumstances and available remedies;
- learn and be upgraded by the experts.

---

[40] See Article 26 Vienna Convention, note 27.

[41] See Article 61 Vienna Convention, note 27; and ICJ Reports, 1973, pp. 3, 20–1.

[42] Charles G. Brown, The Doctrine of Impossibility of Performance and the Foreseeability Test, 6 Loy. U. Chi. L. J. 575 (1975). Available at: http://lawecommons.luc.edu/luclj/vol6/iss3/4.

[43] Offer was made within the Program PRISMA of the Science Fund of the Republic of Serbia https://fondzanauku.gov.rs/programs/?lang=en.

The software was intended to be trained by legal experts and implemented at the Court of Appeals in Kragujevac.[44] The software price was estimated at 78.016,95 EUR and a total price ought to be 297.843,26 EUR for three years for its development and implementation. This amount could not be qualified as an excessive and unreasonable cost either from an objective or from subjective perspective. Therefore, since the cost does not fall within the scope of the present criterion it appears that the doctrine of impossibility of performance cannot apply and give relief to the obligation to use AI.

In the normative sense, reflections of the doctrine of impossibility of performance could be found at the 1969 Vienna Convention. In Article 61, the 1969 Vienna Convention stipulates that the impossibility of performing a treaty as a ground for terminating or withdrawing from it might be invoked by a party if the impossibility results from the permanent disappearance or destruction of an object indispensable for the execution of the treaty.[45] Temporary impossibility was further recognized as a ground for suspending the operation of the treaty while breach of an obligation was precluded from excusing grounds for terminating, withdrawing from, or suspending the operation of a treaty. The relevance of this Article for the eventual obligation to use AI is troublesome in several aspects.

First, Article 61 refers to the supervening situations – those that commence at the time when the treaty was entered into. In that scenario, purposed AI should initially exist - at the time of the contract conclusion, and subsequently get permanently destroyed as a result of unforeseen occurrences which is not the case here. Second, purposed AI should be an indispensable object for the treaty execution. To clarify this requirement, we should understand what is deemed to be an indispensable object. It is a narrowly defined concept comparable to the river at which fishing was contracted. If the river dries up impossibility of performance of the fishing contract applies.[46] At the moment, missing AI is not that important. Rather, it should be understood as a tool for treaty performance, or a specific performance, not as an object for its execution.

Another (quasi)normative reflection of the doctrine concerned could be found in the Principles of European Contract Law (PECL).[47] As a set of model rules that aim to clarify the basic rules of contract law and the law of obligations common to most legal systems of the European Union member states, the PECL is not a typical source of international public law nor communitarian law. It serves as a model for judicial and legislative development of contract law and a basis for harmonizing the Member States' contract laws.[48] As such it is influential on the Member States and important to Serbia in the context of legislation harmonization. For those reasons, insight into its optics on

---

[44] Agreement signed between the Faculty of Law in Kragujevac and the Court of Appeals in Kragujevac on 13th November 2021.

[45] Article 61 Vienna Convention, note 27.

[46] See Giegerich, T. (2018). Article 61. In: Dörr, O., Schmalenbach, K. (eds) Vienna Convention on the Law of Treaties. Springer, Berlin, Heidelberg. https://doi.org/10.1007/978-3-662-55160-8_64.

[47] Principles of European Contract Law – PECL. Available at https://www.trans-lex.org/400200/_/pecl/#head_123.

[48] Dionysios P. Flambouras, The Doctrines of Impossibility of Performance and Clausula Rebus SIC Stantibus in the 1980 Convention on Contracts for the International Sale of Goods and the Principles of European Contract Law - A Comparative Analysis, 13 Pace Int'l L. Rev. 261 (2001)

the issue could be useful. In Article 8:108 the PECL introduces an excuse due to an impediment with the following elements: non-performance results from an impediment beyond the control of the non-performing party; a particular impediment could not be reasonably foreseen (at the time of the conclusion of the contract), the non-performing party could not avoid or overcome the impediment or its consequences.

If those elements are proven to exist then the non-performing party is relieved from the performance of specific obligations (as outlined in PECL Article 9:102), or the cancellation of the contract.[49] If we neglect formal obstacles to the PECL applicability to Serbia – EU contract, then we should see if the non-application of purposed AI could be qualified as an impediment and if so, can it be excused under the PECL? As we have seen laid down elements of the impediment are much similar to that of force majeure.[50] In Article 1218 of the French Code Civil impossibility of performance and force majeure are synonymous even though they should not be confused.[51] For this reason, the non-application of purposed AI cannot be subsumed/excused under Article 8:108 of the PECL.

Now, we should analyse if there is an actual obligation to use AI. At the beginning of this subsection, we have determined that Serbian obligations at the Accession Negotiation are that of the result. Further, we saw that Serbia is underperforming its contractual obligations and AI could accelerate its advancement.[52] Also, we saw that Serbia cannot be excused due to an impediment. Still, is all of this enough to claim for legal obligation on the part of Serbia to use the purposed AI? A negative answer could be supported by the fact that the use of purposed AI cannot be found in the wording of the Accession Negotiation relevant instruments. In accordance to the textual approach, the primary role is given to the wording in the interpretation of a treaty because the wording is regarded as an authentic representation of the parties' intentions.[53] This was followed by Article 31, paragraph 1 of the 1969 Vienna Convention. The textual or formalistic approach to treaty interpretation can be further traced through the ICJ's case law.[54] However, the textual approach is not absolute. Even though it dominates in the field of treaty interpretation, it

DOI: https://doi.org/10.58948/2331-3536.1212 Available at: https://digitalcommons.pace.edu/pilr/vol13/iss2/2.

[49] Ibid.

[50] Katsivela, M. (2007). Contracts: Force Majeure Concept or Force Majeure Clauses. Unif. L. Rev. ns, 12, 101.

[51] See Jovičić, K., Јовичић, К., Vukadinović, S. и Вукадиновић, С., 2023. Neizvršenje ugovora, odgovornost i naknada štete. Beograd: Institut za uporedno pravo. pp 106.

[52] Dragan Dakic, Artificial Intelligence as an accelerator of European Union Accession 2stSerbian International Conference on Applied Artificial Intelligence (SICAAI) Kragujevac, Serbia, May 19–20, 2023.

[53] See Jean Salmon at Laurence Boisson de Chazournes, Anne-Marie La Rosa and Makane Moïse Mbengue in Olivier Corten and Pierre Klein (eds) The Vienna Convention on the Law of Treaties: A Commentary (OUP 2011), Art 26.

[54] Legality of Use of Force (Yugoslavia v. Belgium), Preliminary Objections, Judgment, ICJ Reports 2004, para 100 cf to Separate Opinion of Judge Kreća.

is not so when it comes to treaty implementation. At the implementing phase, the prevalence is rather given to the purpose of the treaty, and the intentions of the parties.[55] This could be confirmed in ICJ's case Gabčíkovo-Nagymaros Project where it was found that "it is the purpose of the treaty, and the intentions of the parties in concluding it, which should prevail over its literal application".[56] This interpretation was given to explain what good faith performance implies.

Since this inquiry concerns the performance of the Accession Negotiation obligations, the abovementioned principle of *bona fides* or good faith should be addressed in more detail. In broader terms, this principle is relevant for both the creation and execution of contractual obligations. The ICJ confirmed the relevance of the principle of good faith in the negotiating phase through its case law.[57] The meaning of this principle at the negotiating stage could be well captured in the Court's statement in the advisory opinion on the Legality of the Threat or Use of Nuclear Weapons which defines the 'purpose of negotiations in good faith':[58]

> The legal import of that obligation goes beyond that of a mere obligation of conduct: the obligation involved here is an obligation to achieve a precise result – nuclear disarmament in all its aspects – by adopting a particular course of conduct, namely, the pursuit of negotiations on the matter in good faith

Therefore, this principle applies before the entry into force of the treaty (tacit *pactum*) which makes it applicable to non-formal agreements as well.[59] However, it should be noted that the principle alone does not possess the capacity to generate enforceable responsibilities in the absence of any pre-existing obligations.[60] In this regard, Serbia has the obligations of the result in priority areas, among others to decrease incompliance of judicial/administrative decisions to the EcHR and the number of domestic cases pending before the ECtHR. The good faith performance requires parties to undertake activities in compliance not just with the letters of the legal obligations outlined in the Accession Negotiation, but also with the spirit[61] and the purpose[62] of the process. More precisely, Serbia should undertake all necessary actions which are not excessive or unreasonably

[55] See Schmalenbach, K. (2018). Article 27. In: Dörr, O., Schmalenbach, K. (eds) Vienna Convention on the Law of Treaties. Springer, Berlin, Heidelberg. https://doi.org/10.1007/978-3-662-55160-8_30.

[56] ICJ Gabčíkovo-Nagymaros [1997] ICJ Rep 7, para 142. Retrieved from Schmalenbach, K. (2018). Article 27. In: Dörr, O., Schmalenbach, K. (eds) Vienna Convention on the Law of Treaties. Springer, Berlin, Heidelberg. https://doi.org/10.1007/978-3-662-55160-8_30.

[57] Cameroon v. Nigeria, ICJ Reports, 2002, pp. 303, 423.

[58] ICJ Reports, 1969, pp. 3, 47.

[59] Reinhold, Steven. "Good faith in international law." *UCLJLJ* 2 (2013): 40.

[60] Nuclear Tests Case, ICJ Reports 1974.

[61] See ICJ Nicaragua (Merits) [1986] ICJ Rep 14, para 270. Retrieved from Schmalenbach, K. (2018). Article 27. In: Dörr, O., Schmalenbach, K. (eds) Vienna Convention on the Law of Treaties. Springer, Berlin, Heidelberg. https://doi.org/10.1007/978-3-662-55160-8_30.

[62] ICJ Gabčíkovo-Nagymaros [1997] ICJ Rep 7, para 142. Retrieved from Schmalenbach, K. (2018). Article 27. In: Dörr, O., Schmalenbach, K. (eds) Vienna Convention on the Law of Treaties. Springer, Berlin, Heidelberg. https://doi.org/10.1007/978-3-662-55160-8_30.

expensive, to achieve agreed results in order to access the EU. At this point, it looks hard to exclude the application of purposed AI from the context of good faith performance.

7 Conclusion

Our research indicates that the obligations of results during the Accession Negotiation are essential. Additionally, we have observed that Serbia is not meeting its agreed obligations, and the use of AI could potentially expedite its progress. Furthermore, we have determined that Serbia cannot use any hindrance as an excuse. However, the claim of duty to use AI in the EU accession process is faced with a lack of its textual foundation. Analysing relevant case law in treaty interpretation has revealed that a strict textual approach is not always definitive. Although this approach is dominant in the field of treaty interpretation, it does not hold the same weight in treaty implementation. During the implementation phase, greater importance is given to the purpose of the treaty and the intentions of the parties involved. Since our investigation focuses on the fulfilment of the Accession Negotiation obligations, we have thoroughly addressed the principle of good faith or bona fides. We have discovered that this principle applies even before the treaty takes effect (tacit pactum) and is also applicable to non-formal agreements. However, we have observed that the principle alone cannot create enforceable responsibilities in the absence of pre-existing obligations. In light of this, Serbia has prioritized obligations in key areas, including reducing non-compliance with judicial/administrative decisions in the ECtHR and decreasing the number of pending domestic cases before the ECtHR. The effective application of good faith requires parties to adhere not only to the letter of the legal obligations outlined in the Accession Negotiation but also to the overall spirit and purpose of the process. Specifically, Serbia must take all necessary actions that are reasonable and financially viable in order to achieve the agreed-upon results for EU accession. Considering these factors, it is becoming apparent that the utilization of AI in pursuit of these objectives cannot be discounted within the framework of good faith performance.

References

1. 12/86, Meryem Demirel v. Stadt Schwabisch Gmünd (ECR, 1987, p. 3719), 192/89, Sevince v. Staatssecretaris van Justicie, (ECR I, 1990, p. 3641)
2. 237/91, Kus, ECR I, p. 6781 (1992)
3. 355/93, Eroglu v. Land Baden Wurrtenberg, ECR I, p. 5113 (1994)
4. 432/92, The Qeen v. Minister of Agriculture, Fisheries and Food, ECR I, p. 3087 (1994)
5. Accession criteria (Copenhagen criteria). https://eur-lex.europa.eu/EN/legal-content/glo ssary/accession-criteria-copenhagen-criteria.html
6. Action Plans
7. Agreement signed between the Faculty of Law in Kragujevac and the Court of Appeals in Kragujevac on 13th November 2021
8. Cameroon v. Nigeria, ICJ Reports, vol. 423, p. 303 (2002)
9. Brown, C.G.: The Doctrine of Impossibility of Performance and the Foreseeability Test, 6 Loy. U. Chi. L. J. 575 (1975). http://lawecommons.luc.edu/luclj/vol6/iss3/4

10. Consolidated version of the Treaty on European Union. https://www.legislation.gov.uk/eut/teu/article/2
11. Constitution of Republic of Serbia. http://www.parlament.gov.rs/upload/documents/Constitution_%20of_Serbia_pdf.pdf
12. Constitution of Republic of Serbia. http://www.parlament.gov.rs/upload/documents/Constitution_%20of_Serbia_pdf.pdf
13. Dakić, D.: Reproductive autonomy conformity assessment of purposed AI system. In: Filipovic, N. (ed.) Applied Artificial Intelligence: Medicine, Biology, Chemistry, Financial, Games, Engineering. AAI 2022, LNNS, vol. 659, pp. 45–57. Springer, Cham (2023). https://doi.org/10.1007/978-3-031-29717-5_3
14. Dionysios, P.F.: The doctrines of impossibility of performance and Clausula rebus SIC Stantibus in the 1980 convention on contracts for the international sale of goods and the principles of European contract law - a comparative analysis. 13, 261 (2001). https://doi.org/10.58948/2331-3536.1212 https://digitalcommons.pace.edu/pilr/vol13/iss2/2
15. Dakic, D.: Artificial intelligence as an accelerator of European Union Accession 2stSerbian International Conference on Applied Artificial Intelligence (SICAAI) Kragujevac, Serbia, 19–20 May2023
16. European commission for democracy through law (Venice Commission), opinion on the constitution of Serbia, No. 405/2006, CDL – AD, 2007 para 14 – 18. https://www.venice.coe.int/webforms/documents/default.aspx?pdffile=CDL-AD(2007)004-e
17. Giegerich, T.: Article 61. In: Dörr, O., Schmalenbach, K. (eds.) Vienna Convention on the Law of Treaties. Springer, Berlin (2018). https://doi.org/10.1007/978-3-662-55160-8_64
18. ICJ Reports, vol. 47, p. 3 (1969)
19. Jean Salmon at Laurence Boisson de Chazournes, Anne-Marie La Rosa and Makane Moïse Mbengue in Olivier Corten and Pierre Klein (eds.) The Vienna Convention on the Law of Treaties: A Commentary (OUP 2011), Art 26
20. Jovičić, K., Јовичић, К., Vukadinović, S. и Вукадиновић, С.: Neizvršenje ugovora, odgovornost i naknada štete. Institut za uporedno pravo, Beograd, p. 106 (2023)
21. Judgment of the Court of 9 March 1978. Amministrazione delle Finanze dello Stato v Simmenthal SpA. Reference for a preliminary ruling: Pretura di Susa - Italy. Discarding by the national court of a law contrary to Community law. Case 106/77
22. Katsivela, M.: Contracts: force majeure concept or force majeure clauses. Unif. L. Rev. ns 12, 101 (2007)
23. Legality of use of force (Yugoslavia v. Belgium), Preliminary Objections, Judgment, ICJ Reports 2004, para 100 cf to Separate Opinion of Judge Kreća
24. Miščević, T., и Мишчевић, Т.: Legislative obligations of Serbia in the accession process to the European Union - the case of fundamental rights. Yearbook. Provincial Protector of Citizens - Ombudsman; Institute of Criminological and Sociological Research.стр, pp. 333–348 (2018)
25. Nuclear Tests Case, ICJ Reports (1974)
26. Official Gazette of the Republic of Serbia, No. 109/2007, 99/2011, 18/2013 - CC decision, 103/2015, 40/2015 - other laws, 10/2023 and 92/2023
27. Permanent Court of International Justice, Jurisdiction of the Courts of Danzing, Ser. B, vol. 15, p. 18 (1928)
28. Permanent Court of International Justice, Ser. A/B, No. 44, p. 24
29. Permanent Court of International Justice, Ser. B., No. 17, pp. 32–33
30. Principles of European Contract Law – PECL. https://www.trans-lex.org/400200/_/pecl/#head_123
31. Reinhold, S.: Good faith in international law. UCLJLJ 2, 40 (2013)
32. Schmalenbach, K.: Article 27. In: Dörr, O., Schmalenbach, K. (eds.) Vienna Convention on the Law of Treaties. Springer, Berlin (2018). https://doi.org/10.1007/978-3-662-55160-8_30

33. Serbia report 2023. https://neighbourhood-enlargement.ec.europa.eu/serbia-report-2023_en
34. Shaw, M.N.: International Law, 9th edn. Cambridge University Press, Cambridge (2021). https://doi.org/10.1017/9781108774802
35. Stabilisation and Association Process - European Commission (europa.eu)
36. Statute of the International Court of Justice. https://www.icj-cij.org/index.php/statute
37. Text adopted by the Commission at its fifty-third session, in 2001, and submitted to the General Assembly as a part of the Commission's report covering the work of that session. The report, which also contains commentaries on the draft articles, appears in Yearbook of the International Law Commission, 2001, vol. II (Part Two). Text reproduced as it appears in the annex to General Assembly resolution 56/83 of 12 December 2001, and corrected by document A/56/49(Vol. I)/Corr.4. https://legal.un.org/ilc/texts/instruments/english/draft_articles/9_6_2001.pdf
38. The Opening Statement of the Republic of Serbia, Intergovernmental Conference on the Accession of the Republic of Serbia to the EU, 21 January 2014
39. The SAA CELEX:22013A1018(01):EN:TXT.pdf (europa.eu)
40. Vienna Convention on the Law of Treaties 1969. https://legal.un.org/ilc/texts/instruments/english/conventions/1_1_1969.pdf
41. Vienna Convention on the Law of Treaties 1969. https://legal.un.org/ilc/texts/instruments/english/conventions/1_1_1969.pdf
42. Vukadinović, R.D., Вукадиновић, Р.Д., Vukadinović Marković, J., и Вукадиновић Марковић, J.: Uvod u institucije i pravo Evropske unije. Beograd: Službeni glasnik.str.402 (2020)
43. Дакић, Д. и Dakić, D.: Аватари као пружаоци услуга: међународноправни аспекти. [на Интернету] Правна регулатива услуга у националним законодавствима и праву Европске Уније: [зборник реферата по позиву са Међународног научног скупа одржаног 19. и 20. маја 2023. године, на Правном факултету у Крагујевцу у организацији Института за правне и друштвене науке Правног факултета Универзитета у Крагујевцу]. Правни факултет Универзитета, Институт за правне и друштвене науке.стр.‡стр. ‡399–412 (2023). http://institut.jura.kg.ac.rs/images/Projekti/7%20XXI%20vek/Pravna%20regulativa%20usluga%20u%20nacionalnim%20zakonodavstvima%20i%20pravu%20EU%202023.pdf
44. Закон о потврђивању Споразума о стабилизацији и придруживању између европских заједница и њихових држава чланица, са једне стране, и Републике Србије, са друге стране (Службени гласник РС – Међународни уговори, бр. 83/2008)
45. Крећа, М. и Kreća, M.: Међународно јавно право. 14. допуњено изд. изд. Београд: Правни факултет Универзитета, Центар за издаваштво.стр.844 стр (2023)
46. Милисављевић, Б., Палевић, М. и Palević, M., 2017. Однос међународног и унутрашњег права према Уставу Републике Србије из. године. Српска политичка мисао, [на Интернету] (издање), стр.Стр. pp. 29–46 (2006). http://www.spmbeograd.rs/arhiva/SPM-2017-posebno/CD%20SPM%20SPC%202017.pdf
47. Радивојевић, З. и Radivojević, Z.: Споразуми о стабилизацији и придруживању у праву Европске уније. Zbornik radova Pravnog fakulteta u Nišu, [на Интернету] (62), стр.стр. pp. 53–72 (2012). http://www.prafak.ni.ac.rs/files/zbornik/sadrzaj/zbornici/z62/04z62.pdf
48. Радивојевић, З., Radivojević, Z.: Споразуми о стабилизацији и придруживању у праву Европске уније. Zbornik radova Pravnog fakulteta u Nišu, 62, стр. pp. 53–72 (2012). http://www.prafak.ni.ac.rs/files/zbornik/sadrzaj/zbornici/z62/04z62.pdf

Author Index

N. Filipović (Ed.): AAI 2023, LNNS 999, pp. 249–250, 2024.
https://doi.org/10.1007/978-3-031-60840-7

Printed in the United States
by Baker & Taylor Publisher Services